COWBOY WRECKS & RATTLESNAKES
Tales of a Texas Brush Country Rancher

by Jack Kingsbery

Copyright 1998 by Jack Kingsbery

First edition published in 1998 by
Second printing in 2000
Kingsbery Communications
Mill Creek, Washington USA

All rights reserved. No part of this book may be reproduced in any form or by any means without written permission from the publisher.

ISBN 0-9667590-0-1

For additional copies send $17.95, plus $2 for postage & handling, to…

Jack Kingsbery
PO Box 477
Crystal City, Texas 78839

Printed in the United States of America

To my good friend Clem
Best Wishes
Jack Kingsbery

CONTENTS

COWBOY WRECKS．．．．．．．．．．．．．．．．．．．．．．．．．．．．．．．．．．．	1
WHERE'S THE MILK?．．．．．．．．．．．．．．．．．．．．．．．．．．．．．．．	5
GROWING UP．．	11
SANCHO & DAINTY．．．．．．．．．．．．．．．．．．．．．．．．．．．．．．．．．	17
THE BEST HORSE I EVER OWNED．．．．．．．．．．．．．	21
MAMA CATS AND BABY SQUIRRELS．．．．．．．．．．．	27
THE OLD BUGGY．．．．．．．．．．．．．．．．．．．．．．．．．．．．．．．．．．．．	31
ROPING STORIES．．．．．．．．．．．．．．．．．．．．．．．．．．．．．．．．．．．	35
MAD COWS & BAD BULLS．．．．．．．．．．．．．．．．．．．．．．．．	39
A TRUE TWISTER TALE．．．．．．．．．．．．．．．．．．．．．．．．．．．	45
THE WILD BUNCH．．．．．．．．．．．．．．．．．．．．．．．．．．．．．．．．．．．	49
THE BLIND COWBOY．．．．．．．．．．．．．．．．．．．．．．．．．．．．．．．	53
WHO GOT RATTLED?．．．．．．．．．．．．．．．．．．．．．．．．．．．．．．．	57
THE WORLD'S SLOWEST HORSE RACE．．．．．．．	63
BUFORD THE JAVELINA．．．．．．．．．．．．．．．．．．．．．．．．．．．	67
COWBOY WRECKS Part 2．．．．．．．．．．．．．．．．．．．．．．．．	69
BICYCLE WRECKS．．．．．．．．．．．．．．．．．．．．．．．．．．．．．．．．	75
THE CLOTHESLINE WRECK．．．．．．．．．．．．．．．．．．．．．．．	79
OIL FIELD TALES．．．．．．．．．．．．．．．．．．．．．．．．．．．．．．．．．．．	81
SQUEEZE CHUTES, SADDLE RACKS & GUN SAFES．．．．．．．．．．．．．．．．．．．．．．．．．．．．．．．．．．．	87
HORSE TALES．．	95
COUGARS, COYOTES & MOTHER HENS．．．．．．．．．	101
MESQUITE CORRAL COWBOYS．．．．．．．．．．．．．．．．．．	105
WILD STEERS IN A GARAGE．．．．．．．．．．．．．．．．．．．．	109
MAURICE AND THE RATTLESNAKE．．．．．．．．．．．．．．	113
LOVE, WAR & CATTLE PRICES．．．．．．．．．．．．．．．．．．	119
THE KILLER BULL．．．．．．．．．．．．．．．．．．．．．．．．．．．．．．．．．．．	123
ALWAYS CLOSE THE GATE．．．．．．．．．．．．．．．．．．．．．．．	125
COWBOYS AND COMPASSES．．．．．．．．．．．．．．．．．．．．．	129
THE BEST COWDOG I EVER OWNED．．．．．．．．．．．	135
ABOUT THE AUTHOR．．．．．．．．．．．．．．．．．．．．．．．．．．．．．．	145

FOREWORD

    I grew up eagerly listening to my father's stories about ranching, hunting and rodeos. Our bedtime stories were not about tales of princesses and fairy godmothers in make believe kingdoms, but true stories about cowdogs, coyotes and, of course, the ever popular "Maurice and the Rattlesnake." Roping a deer or riding a runaway buggy seemed much more interesting and believable than tales starting out "Once upon a time..."
    Over the years I began to realize how rich and unique my father's life has been. Very few people have experienced as much as Jack Kingsbery. Indeed, his stories about cowboy wrecks make us wonder how he survived at all. Now that helicopters and dart guns are used to round up cattle, Jack reminds of us an era when cowboys caught cattle the old-fashioned way, on a horse running full-tilt through the thick brush of South Texas.
    My city friends enjoyed my tales about growing up on a ranch; catching a javelina and chasing wounded deer. I knew they would be fascinated by my father's much more interesting stories. For years we urged him to write them down. He put it off until he sold his company in 1997. I remember sitting down with him to make a list and we came up with 14 stories. Once he started writing the memories began to flow. He kept sending me the "last" story, only to think of another one. Fifty plus stories and 30 chapters later this book came into being.
    I enjoyed reading the stories I had heard as a child, plus some he never got around to telling us. I am sure you will also. You can write Jack at PO Box 477, Crystal City, Texas 78839.

Bob Kingsbery
Mill Creek, Washington

DEDICATION

To all the cowboys, cow horses and cowdogs that made my life such a wonderful experience. Not to mention the rattlesnakes and javelinas. Without them I wouldn't have any stories to tell.

Also to my wife, Evelyn, for putting up with all my wrecks, ranching and rodeos. Having her beside me is the best story of all.

SPECIAL THANKS

To my son Bob, ranch-raised and a writer in his own right. For editing and publishing this book with his wife, Kerri. Also to Lois Youngman, for proof reading my stories so carefully.

And to Cynthia Buchanan, writer and teacher, and all the members of our writing class. For their encouragement and enthusiastic response as I was writing these stories.

Without their help this book would not have seen the light of day.

COWBOY WRECKS

I grew up on a ranch south of Santa Anna in Coleman County, Texas with two brothers. Tom was the youngest and Hank was the oldest. Tom and I did most of the cowboy work--breaking colts and doctoring screw worms in our cattle and sheep. For fun we often rode the milk pen calves after milking the cows. Dad never objected to us riding the calves after we finished all the evening chores.

One evening, when I was about five or six years old, I was riding one of the bigger calves, holding on to a rope around the calf's girth. The calf ran through a barbed wire fence and the barbed wire cut a deep gash above my right ankle. The cut was about three inches long and all the way to the bone. My folks took me to the hospital to get the cut sewed up. Santa Anna had a big hospital, which was unusual for such a small town. Dr. Sealy was the local doctor and since it was after dark we drove to his house. He told Dad he would meet us at the hospital.

My leg was really hurting but I wouldn't complain because I was afraid Dad would make us stop riding the calves when he saw how bad my leg was cut. Dr. Sealy didn't give me anything for the pain as he was sewing up the cut. The stitches made the cut hurt a lot more, but I still wouldn't cry or complain. Dr. Sealy asked me if I was going to ride any more calves and I told him that I was. Dad got nauseated when Dr. Sealy started sewing my cut and had to go outside and sit on the hospital steps. Of course our mother didn't want Tom or me to ride calves after that, but Daddy never told us not to ride them. My cut healed real fast and in a month or two we were back riding the calves again.

The larger calves could really pitch hard and it took lots of skill to ride them. Tom and I often talked about riding the yearling cattle with a saddle. When Tom and I were twelve and thirteen years old we were riding two good horses, checking cattle in one of the pastures. There was a big yearling heifer in the herd with horns about six inches long. The pasture didn't have many trees to worry about so I decided to try saddling the heifer for a ride. Tom and I both roped her and Tom held her with his rope while I tied my rope to a mesquite tree. When she was stretched between the tree and Tom's horse he got off and held her by the head and horns. I took the saddle off my horse, put it on the heifer, tightened the girth and got on.

As Tom was taking the ropes off the heifer I spurred her and off she went, bucking high and hard. Unfortunately she jumped before Tom got the rope off the tree. When she hit the end of the rope she turned a flip and landed on her back with me underneath. The heifer jumped up and continued bucking around the tree. I hurt so bad I thought I was going to die.

Tom caught the heifer, took the saddle off and turned her loose. When he saw I was going to live he saddled my horse and helped me climb on. We agreed not to tell Dad what happened. He didn't mind us riding and roping the milk pen calves but didn't want us playing around when we were supposed to be working.

My hip hurt so bad I though it was broken. I kept on doing the work on the ranch even though the pain was terrible. The pain gradually went away and was gone in three or four months. Years later when I took a physical to get into the Texas A&M Corps of Cadets I found out my right hip had been knocked down by the impact. To this day I have to get the right leg altered on my pants so they fit.

⬌

In 1941 I entered Texas A&M College with nineteen hundred eighty-five other students. In December of that year the Japanese bombed Pearl Harbor and America was at war. At the end of our sophomore year every member of our class went into the military services. I joined the Army Air Corps (now the U.S. Air Force) and served in England in the Eighth Air Force for nineteen months.

I was discharged in October 1945, too late to return to Texas A&M for the fall semester, so I went home to the ranch in Coleman County. I helped Dad on the ranch that fall doing things that had gone undone while my brothers and I were in the service for three years. Around Christmas I started breaking a big, stout three-year-old mare we had raised on the ranch. I had her going real good and needed to get lots of miles on her before I went back to A&M for the spring semester. By riding her hard for my last two weeks I figured Dad could start using her when I left.

Early one morning I told Dad I was going to ride the mare a long way on the ranch and have her going good by that afternoon. It was a warm morning so I wore a light brush jacket and rode off on the young mare about eight o'clock. I had ridden her about two miles when a big jackrabbit jumped up right in front of her. The mare snorted and started pitching high and hard in a circle. It had rained some the night before and the ground was a little slick. I was managing to hang on when all of a sudden the mare's front feet slipped out from under her and she went down on her right side. I threw my right leg up as quickly as possible but was too late. The full weight of the mare landed on my right leg. I heard the bone in my leg pop like a shotgun going off.

I managed to catch the reins of the hackamore and thought I could hold the mare down while I got my foot out of the stirrup. The mare was too strong. She jerked the rein out of my hand and got up. My right foot was caught in the stirrup and the mare started pitching and kicking, dragging me underneath her. There was quite a bit of brush in the pasture and the mare was dragging me through a lot of it, kicking me on the side and shoulder at the same time. Once I was able to grab hold of the stirrup to try to get my foot out, but we hit some brush and it knocked me back under the fast moving mare.

About that time the mare headed for a big mesquite tree with a straight trunk and no lower branches. She started to the right of the tree but suddenly ducked to the left of the tree. The sudden turn swung me out from under the mare and the stirrup leather partially wrapped around the trunk of the tree. My foot wedged against the tree, causing the stirrup leather to pull completely out of the saddle. I hit the ground hard, lucky to be alive.

I knew my leg was broken and when I looked there was blood coming through my Levis just below my knee. I carefully pulled off my boot and pulled up my pants leg. The bone was sticking out through the skin and bleeding bad. I knew I needed to slow the bleeding so I cut off a strip of shirt and tied it around my leg just above the knee. I used my knife to twist the tourniquet tight and just about stopped the bleeding.

It was still early in the morning and I knew Dad wouldn't start looking for me until after dark because I told him I would be gone all day. A cold norther' blew in soon after the wreck and it started raining. I was wearing only a light jacket so I got really cold as the day wore on. The whole thing happened about a hundred yards from a lane that ran along the back of the ranch. One of our neighbors, Carl Williams, had some cattle on a ranch below us and he checked on them once a week. Fortunately it just happened to be the day he drove down to see his cattle. About five o'clock Mr. Williams drove by and I waved my hat to get his attention. He stopped to see what was wrong and I told him to go to our house and have Dad call the ambulance. I knew it would take a stretcher to get me over the net wire fence.

About half an hour later the Hosch brothers, Pat and Norman, came in their ambulance, lifted me over the fence and took me to the hospital in Santa Anna. The X-rays showed the leg was broken in three places. By the next morning I had pneumonia from being out in the cold, wet weather all day. Because of the pneumonia our family physician, Dr. McDonald, couldn't give me ether to set my leg for about a week. Once the pneumonia passed he set my leg using two steel plates screwed to the bone and put it in a cast. My leg was in a full length cast for about three months, then in a walking cast for another three months. As a result of the wreck I missed the spring semester at Texas A&M.

There was a roping arena in Santa Anna where the local cowboys roped goats every Sunday afternoon. I was anxious to start roping again so I rigged a leather strap as a stirrup for my walking cast. Two of my friends, Willard and James Allen, were always there. James was only nine years old and, since I couldn't get off my horse when I roped, he would race to the goat on his little Shetland pony and practice throwing and tying it. James grew up to become a top professional steer roper. His son, Guy, has been a World Champion steer roper several times.

Later that summer Dr. McDonald took the steel plates off and I started back to A&M in the fall. I guess Dr. McDonald did a good job because I started roping calves in rodeos later that fall and haven't had any problems with the leg since.

WHERE'S THE MILK?

While attending Texas A&M in the late 1940s a bunch of us Aggie cowboys would participate in rodeos in the area. Back in those days one of the most exciting events was the wild cow-milking contest. It was a timed event and a real crowd-pleaser. The contestant would have a partner, called a "mugger," in the arena, either on foot or on horseback. The cows to be milked were young Brahman or Brahman-cross that usually had big horns and were really wild. The cow would be turned out of the chute and the roper would rope the cow by the horns or head, dally his rope around the saddle horn and stop the cow. The mugger would grab the cow by the head and hold her.

The roper carried an empty Coke bottle in his hip pocket and when his mugger got a good hold on the cow, he would dismount and run to the cow. He would reach down under the cow, grab a teat and try to get one or two squirts of milk in the Coke bottle. He would then run back to the finish line with his bottle of milk. The time for the event began when the roper started after the cow and stopped when the roper crossed the finish line. There had to be enough milk in the bottle so it would drip out when the bottle was turned upside down.

On top of that, the roper couldn't cross the finish line until his partner had taken the rope off the cow. The cows were really wild and that could be a chore. Several times I have seen cows get the mugger down on the ground before he could get the rope off.

One time I was entered in the wild cow milking at Caldwell, Texas. I remember they had the rankest cows I had ever seen used in that event. They were too big to go through the dogging chute so they had to open two bucking chute gates and tie them together to hold the cow to be roped. I was the fourth roper and the muggers on the first three cows had really gotten beat up. As I rode into the roping chute to get ready my cow was trying to jump over the top of the bucking chute gates. One of the chute men had to push her off the gates to keep her from jumping over. She had big horns and came out of the chute running like a deer. Just as I went by my partner, who was standing in the middle of the arena, hollered as loud as he could, "Miss her Jack, miss her!" Unfortunately for him I caught her and he got in such a wreck trying to mug her that we didn't' place in the contest.

One weekend we went to a rodeo in Cameron, Texas. One of our fellow calf ropers was Bill Soyars of Sabinal, Texas. He lives in San Marcos now and was one of the best ropers at A&M. His horse, Frankie, was about the best roping horse in the rodeo, with excellent speed and a great stop when roping calves. Bill was real friendly and liked by every cowboy. He also liked to kid the other cowboys. He was the sixth or seventh roper in the wild cow-milking contest and at the rodeo there was a dogging steer that had a feminine head. Some of the cowboys from A&M paid the man who was putting the cows in the chute five dollars to put that steer in the chute when Bill's time came up in the cow milking.

Bill was riding Frankie and they made a good run, roping his "cow" right out of the chute. His mugger grabbed "her" head and Bill turned his dally loose, bailed off Frankie, grabbed the Coke bottle out of his hip pocket and ran to his "cow." Bill reached under the "cow" and kept feeling around for a teat. With a puzzled look on his face he kneeled down and looked under the "cow." Then he got up, threw his Coke bottle on the ground and got back on Frankie. We were all laughing so hard we could hardly stand up. None of the Aggies would admit responsibility for the exchange and Bill still thinks I was the leader in the "mix up." To pour salt on the wound, Bill was given a rerun on a real cow but missed with his loop and got no time in the event.

◀──────▶

One summer a bunch of us Aggies went to Southwest Texas State College in San Marcos to take courses that we could transfer back to A&M the following fall. The courses at San Marcos were about as hard as the courses at A&M but SW Texas had two things that made it better. Girls were the number one reason (at the time Texas A&M was still an all-male school). All the rodeos in the area were a distant second.

One weekend all of us went up to Johnson City to a rodeo where they also had match horse races. A match race is where two horses race against each other. We hadn't done very well at the rodeo on Friday night—I think I caught my calf but was too slow tying it—so we decided to go to the races

Saturday afternoon to see if we could match Bill Soyar's Frankie in a race. We knew Frankie had lots of speed but he never had been raced. Most of us were going to college on the G.I. Bill and none of us had much money. We had used up most of it for the rodeo entry fees and couldn't cover a very big bet on a match race. We pooled our money and only came up with seventeen dollars.

    I knew several people from the local area who told me there was only one top racehorse at the track, a four-year-old bay named My Baby. She was owned by a Mr. Bierschwale and had won her last six match races. We tried to match all the other horses for a race except My Baby but Frankie was such a good-looking horse nobody would race their horse against him. I told Bill that I thought Frankie could outrun My Baby. We approached Mr. Bierschwale but he was skeptical when we told him Frankie had never been raced. He thought Frankie might be a "ringer" brought in from another part of the country so the local horsemen didn't know him.

    Mr. Bierschwale really looked Frankie over, picking up his feet and was surprised to see regular horseshoes instead of aluminum racing shoes. He also felt Frankie's ankles and said they were somewhat sore. We told him three ropers had roped calves off Frankie the night before. Several local people told him they knew the horse was a roping horse and not a trained racing horse. Finally he agreed to race My Baby against Frankie in a 200-yard match race. When we told him we only had seventeen dollars to bet he knew Frankie wasn't a ringer. At first he didn't want to run his mare for such a small amount but since nobody else would race against My Baby he finally agreed.

    We didn't have a jockey or a jockey saddle or a racing bridle—one with snaffle bits. The only person at the track with extra equipment was Mr. Bierschwale and he loaned us a saddle and a bridle and a racing bat. I told Bill I would ride Frankie since I was the lightest cowboy of the bunch. I hadn't gained back all the weight I had lost when I broke my leg and weighed about 145 pounds. Mr. Bierschwale's jockey weighed about 110 pounds but Frankie was so big and stout I didn't think the extra weight would be much of a factor in the race.

    I had never ridden a jockey saddle before but didn't think it would be a problem. Some of the cowboys started looking for a pair of goggles for me to wear but I told them I didn't need them because the mare wasn't going to get in front of me to kick dirt in my face. My main worry was getting Frankie out of the starting gate as fast as the mare. Since Frankie had never been in a starting gate I was afraid he would hesitate when the gate flew open in front of him. My good friend Friday Ellis from San Marcos was working the rodeo with his trick mule that weekend and was at the races. He said he would get behind Frankie and pop him on the rump as the starting gate flew open.

We saddled Frankie with the jockey saddle, put the racing bridle on him and I climbed on as did the other jockey on My Baby. The jockey saddle was sure different from a regular saddle, my knees were right up under my chin and I hoped I wouldn't fall off as Frankie went flying down the track. Since I was tall, skinny and two feet taller than the other jockey, my friends said I looked like Ichabod Crane sitting on Frankie. The track didn't have rails along the sides and the spectators lined up on both sides so we rode Frankie and My Baby down the track to familiarize the two horses with the track and the people. We then rode around the starting gate and into our gates. The track men closed the rear gates and we were ready for the big race. Friday Ellis stood behind Frankie with a small rope to pop Frankie on the rump when the gates flew open.

When our horses were standing just right we both nodded that we were ready. The starter jerked the rope and the gates flew open. Frankie hesitated for just a moment until Friday popped him on the rump and he was off and running. My Baby got about a half-length head start out of the starting gate. I popped Frankie several times on the rump with the bat and by the time we had run about a hundred yards Frankie had caught My Baby and we were running side by side.

About that time we came to where the spectators were lined up on both sides of the track. Something must have spooked My Baby as we neared the crowd because she left the track and ran behind the spectators, parallel to the track. Frankie ran straight down the track and as we crossed the finish line I glanced over at My Baby and we were just about even. She was about 20 yards off the track but still running parallel to it. The race judges had a conference and after a few minutes declared Frankie the winner because he had stayed on the track. We made seventeen dollars. Had My Baby not left the track the race would have been close, I don't know which horse would have won. The cowboys sure kidded me about how funny I looked riding that jockey saddle.

There was a reporter from the ***Austin American Statesman*** newspaper at the races and in his column the next day he stated that our race was the featured race of the meet. He told about My Baby blowing the track and how our horse ran straight and true down the track and crossed the finish line between the two judges. The reporter finished his story by writing that the owner of the "San Marcos Sorrel," as he called Frankie, claimed he had never been raced. The word "claimed" was written in italics.

◄─────────►

In 1948 I was on the Texas A&M rodeo team when we went to an intercollegiate rodeo at Sul Ross State College in Alpine, Texas. There were about fifteen college and university rodeo teams participating in the three-day event that ended Saturday afternoon. A big dance was scheduled for Saturday night at the Turf Club north of Alpine.

Evelyn Bruce, from my hometown of Santa Anna, was a junior at Sul Ross and on the rodeo team. I had dated Evelyn a couple of times when we were both home during college breaks. I called her from College Station for a date on Thursday night, the first day of the rodeo, but she said she already had a date. I asked her about the other nights and about the dance but she said she already had dates. That was all right, I married her in 1951.

It was a good rodeo and our A&M team won the team trophy. Bubba Day and I won second in the steer roping. I hazed for Charlie Rankin and he placed in the bulldogging. I drew a big, rank calf in the calf roping and caught it right out of the chute but had trouble tying it and didn't place. We all went to the dance at the Turf Club that Saturday night and the A&M cowboys were really celebrating after winning the team trophy.

We had our horses stabled at the Turf Club barn and one of our best horses was King Banner. He was a stud sired by King P-234 and was an outstanding dogging and roping horse. The dance had been going on for several hours and most of the cowboys were feeling pretty lively. The place was real crowded and one of my teammates popped off and said, "Jack, why don't you ride King Banner in here on the dance floor?"

I never drank so I was completely sober but for some reason I popped off and said, "If you bring him in here I'll ride him." I didn't think any more about it until the side door opened right by the table where we were all sitting and someone said, "Jack, here he is. Ready for you to ride." There stood King Banner with his head in the door. He had a halter and a lead shank on so I got up from the table, jumped on him and rode him through the door and out on to the dance floor.

The band was playing and lots of people were dancing. King Banner and I made two rounds of the dance floor and the dancing couples would just move over, opening up a lane for us. After the second round the manager came over and said I should ride back outside since somebody might get hurt. King Banner and I got a big hand as we rode off the dance floor and out the side door.

On the way back to College Station the next morning we came upon the Texas A&I College cowboys between Sanderson and Dryden. They had run out of gas so we stopped and started siphoning gas out of our pickup for theirs. One of the A&I cowboys called me around to the front of our pickup and asked me in a low voice so the others couldn't hear, "Jack, was that you riding a horse on the dance floor last night? Or was I so drunk I just imagined I saw you?"

I had to admit it was me riding a horse on the dance floor. He said, "That makes me feel better. I guess I wasn't in as bad shape as I thought."

GROWING UP

My first day at school was different from most kids. My mother, Mabel Kingsbery, was my teacher and my grandfather, H.W. Kingsbery, had donated the land for the Leedy School. It had three rooms with three grades in each room. There were six of us in the first grade and we sat in the first row of seats. The second graders sat in the second row and the third graders behind them.

On the way to school that morning my mother had lectured me about behaving myself in class. During the lunch hour some of the older boys were outside playing marbles for keeps. They let me join their game, probably figuring a first grader would be easy pickings. But I was a pretty good marble player and in a few minutes I had won all their marbles. I was really proud of myself until someone told my mother about me winning all the marbles. She made me give them all back. She told me that was gambling and we weren't supposed to gamble. I guess that made an impression on me because even though I've owned racehorses I never bet on them and have never bought a lottery ticket.

◀——————▶

Mechanical things have always fascinated me. As a child I was always building things such as little cars, tractors and trains. At an early age I got an erector set and could build a lot of things, but it was small. I saved my money and bought two more sets so I could make more and bigger things. One time we were visiting friends in Dallas who had a son my age. He had an erector set and built about the same things I had except he had an electric motor that would make things move. I was fascinated by the motor and got to thinking about how I could power my toys.

We didn't have electricity on the ranch so I couldn't use an electric motor. I had taken alarm clocks apart and put them back together since I was six years old. I knew clocks were run by a big spring that you wound up and by taking the balance wheel out of an old alarm clock that the hands would turn real fast. That would be my motor!

I took the hands off an old clock and attached a pulley wheel to the hour hand shaft. When the clock was wound up the pulley wheel would turn fairly fast. If I wanted more speed I would attach the pulley wheel to the minute hand shaft. I put a small belt on the pulley wheel to run my various contraptions.

I made a little car and a little tractor that would run until the spring wound down. My favorite was a drilling rig. An oil company had drilled oil wells on the ranch and I copied their drilling rigs. I could even drill a small hole in the back yard using the clock motor to run my little rig. I quickly learned to keep all the little wheels in the clock oiled because they ran so much faster than when the clock was keeping time.

I read everything I could find on electric motors—they fascinated me. One of the books my mother got for me told how to make a little electric motor that would run on a nine-volt battery. I read everything in the book on how to make the little motor and started getting the materials that were needed. That was a problem because I was only nine years old and some things were very hard to find. The two main items were a battery and some small-diameter copper wire. You couldn't just go to a store and buy those things like you can now.

The battery wasn't a problem. We had a telephone and I had thoroughly explored it. The phone was the old-fashioned kind that was a large box that hung on the wall. You had to turn a crank to get the operator who connected you to whomever you wanted to call. I had taken the screws out and opened it up many times--when my folks were gone. It had two nine-volt batteries that made it ring—there was my needed battery.

There was no copper wire around the ranch and none to be bought in town. I remembered seeing a coil from an old Model T Ford that someone had taken apart a year or two before. There was a lot of copper wire wrapped around the slate core in the coil. There were still a few Model Ts in the area and after lots of hunting I finally found an old coil and got all the copper wire I needed.

With all the materials collected I went to work making my first electric motor. I wrapped the copper wire around a wooden peg and left the opposite ends sticking out. I drove a small nail in each end of the wooden peg and put it in a wooden frame so it could spin. I had two large wires running parallel above the peg and attached to the battery that I had "borrowed" from our telephone. This created an electromagnet and with just a little adjustment my little motor would run. By reversing the wires attached to the battery I could make it run the other way. I would always put the batteries back in the telephone when I got through playing with the motor.

I made several more motors and before long the batteries ran down and our phone wouldn't ring. Dad saw the telephone company repairman in town and told him our phone didn't work. He came out and put in two new batteries. It wasn't too long before he had to replace those two. About that time I was getting tired of my new project so we didn't have any more phone problems. We got electricity on the ranch in 1935. Then I could use a bigger electric motor and really make things move.

◀——————▶

In the mid-1930s my father, Howard Kingsbery, bought a Farmall F-20 tractor to replace the mules we used for farming in Coleman County, Texas. My two brothers, H.W. and Tom, and I had been hounding Dad to buy a tractor. We didn't do much of the farm work since we weren't even teenagers yet, but we thought farming would be a lot more fun with a tractor. We argued about who got to drive the tractor at first, but it didn't take long for the new to wear off and Dad had to tell which one of us to drive.

The Farmall had iron wheels with steel lugs on the rear wheels. The lugs made the tractor rough to ride and driving it all day was really a chore. After we had owned the tractor about a year the tractor companies introduced rubber tires. They were a big improvement over the jarring ride you got from a tractor with steel lugs and Dad decided to have them installed. To do this the tractor had to be driven to Coleman, about 20 miles from our ranch, so the tractor dealer could replace the steel wheels with rims and tires.

I was twelve years old at the time and Dad told me to drive the tractor into Coleman. To get there I had to drive in the ditch so the lugs wouldn't tear up the pavement. One place where I went down a little hill I put the tractor in neutral gear and let it coast. The hill was fairly steep and the tractor picked up a little speed but the heavy wheels kept it from going too fast. It was a rough ride to Coleman and I was really looking forward to the trip home since I could drive the tractor on the paved highway.

It took the dealer about three hours to install the rubber tires. As I drove out of Coleman heading home the tractor rode so smoothly it was like getting out of a Model T and into a

Cadillac. There was a steep hill, about half a mile from the top to the bottom, on the highway between Coleman and Santa Anna. On a previous trip I had coasted down the hill in neutral in the ditch, just a little faster than normal speed. This time I was enjoying the drive home on rubber tires and when I got to the hill I put the tractor in neutral at the top the hill. Just as I started down a rancher driving a Model A Ford passed me. The tractor started to pick up speed and for a few moments I was enjoying the ride. Then I began to notice that the tractor, with its new rubber tires, was really beginning to move so I thought I had better slow it down.

I reached for the brake lever and pulled. Unfortunately Farmall tractors in those days only had one brake and it was on the left rear wheel. For the last month or so the brake hadn't been working properly. If you pulled the brake lever it would sometimes grab and lock the left wheel. By the time I decided to brake the tractor was going about 25 mile per hour. That doesn't sound very fast, but for a young boy on a big tractor going down a hill it was real fast.

When I pulled the brake lever the brake partially locked and the tractor lunged to the left, almost turning over. It skidded off the highway and into the ditch before the brake unlocked. I managed to straighten it out and drive back on the highway. I realized I couldn't use the brake so I tried to put the tractor back in gear to let the engine slow it down. By then the transmission was turning faster than the engine so I couldn't get it in gear. I gave the engine full throttle, hoping it would let me shift into high gear, but the transmission just made loud grinding noises.

I was going about 40 miles per hour by then and quickly catching up to the Model A that had passed me at the top of the hill. Fortunately there weren't any cars coming toward me so I pulled around and passed the car like it was standing still. I glanced at the rancher and he sure had a surprised look on his face as I went by. I stayed in the middle of the highway and the tractor continued to pick up speed. It seemed like the tractor was flying by the time it reached the bottom of the hill. It coasted for a long time before I could put it back in gear and continue on.

It was a wild and scary ride—I would have felt safer riding a wild bronc. As the rancher passed me farther down the highway he looked at me and just shook his head.

My two brothers and I thought our parents were too strict at times, but looking back I know they were great parents. They steered all three of us in the right direction because we all three turned out OK. One thing that sticks in my mind was that at the age of ten we were allowed to write checks for whatever we needed on Mom and Dads' bank account. All three of us looked forward to that day and honored the privilege by not taking advantage of it. Even though we all three wrote checks for many years Dad never questioned any of the checks we wrote.

SANCHO & DAINTY

One time I bought 150 young crossbred cows and calves from Leroy Hinds. I put them in a small pasture the day they arrived. The next morning while I was checking the cattle I noticed a young calf without a mama. The calf was about two weeks old, three-quarters Brahman and a real pretty light dun with big ears and little hump on his back. I called Leroy and told him about the extra calf and asked him to check and see if he had a cow without a calf at his ranch. He called back later that day and said all his cows were paired up and for me to keep the calf.

Since the calf had lots of Brahman blood I knew he could survive by stealing milk from the other cows. I watched him for awhile and noticed he would wait for another calf to start sucking its mother then he would run up and start sucking the cow at the same time. Most of the cows would let him suck for a while before kicking him off. Then he would go to another cow for a quick meal. I liked his determination and named him Sancho.

He made it through the winter in pretty good shape and in the spring I castrated him along with the rest of the bull calves. Sancho really started putting on weight and by the following spring he was big and fat and ready to ship with my other two-year-old steers.

I had two loads of steers ready for the San Antonio Stockyards and Gene Allen sent two of his big cattle trucks to my ranch to haul them. We had penned Sancho along with the other steers in the water lot near the corrals. As we were driving the herd into the corrals Sancho jumped the fence and trotted down to the nearby Palo Blanco Creek where he had been grazing all winter. It was almost dark by then so I told the other cowboys just to leave him and we would send him with another bunch next month.

A month later I had the truck driver back his truck up to the loading chute with the tailgate open. That way we could push the steers out of the water lot, into the corrals and on the truck real fast, not giving Sancho a chance to jump the fence. Things were going great and I thought we would have Sancho loaded on the truck before he realized what was going on. But just as he got to the loading chute he wheeled and jumped over the high corral fence. Then he just stood there and watched us load the other steers. I figured if he wanted to stay that bad I would just wait and ship him next year.

Sancho was real gentle and you could drive him anywhere. I ran him with the cows and calves since I had sold all my steers. All the next year when we would round up the cattle Sancho would go right into the holding trap along with the cows and calves. He would stay in the trap for a few minutes and then easily hop over the fence and graze around the corrals waiting for us to finish working the calves. Then he would rejoin the cows and calves as we were driving them back to the pasture.

The following spring Sancho was three years old and getting really big and good looking. I decided to load him in my trailer and tie him so he couldn't get out. We rounded up the cows and calves and drove them into the water lot. There was lots of fresh grass in the lot and Sancho was enjoying the grazing along with the cows and calves. He evidently didn't realize that he was so close to the main corral with the corral gate being open. We pushed a bunch of cows and calves into the corral real fast and Sancho found himself trapped before he realized what had happened. I thought he might be too big to jump out of the corral, but over the corral fence he went. Once again Sancho went back to the pasture with the cows and calves.

Later, I had another load of steers to ship to the stockyards and I was determined to get Sancho loaded on the truck. I put some boards around the top of the corral to keep Sancho from jumping out. I told Gene Allen's driver, Walter Nation, to be ready to shut the tailgate on his truck as soon as the last steer was on the truck. The truck had one partition gate in the middle so we put half the steers in the corral, leaving Sancho with the rest of the steers in the water lot. We loaded the first bunch of steers in the front section of the truck then started Sancho's group running from the water lot toward the corrals. We were whooping and hollering and popping our ropes to keep the steers from slowing down. The whole group, including Sancho, ran up the loading chute and into the truck before they knew where they were. Walter slammed the tailgate shut and I breathed a sigh of relief.

It was almost dark by then, so I told Walter to head for San Antonio and not stop until he backed up to the loading docks at the stockyards. He drove off and we started closing the gates around the corral and loading our horses in the trailer.

Just as we were finishing I saw headlights coming back toward the corrals. As the lights got closer I could see it was a big cattle truck. It pulled up in front of the corrals and as I walked up to the cab Walter said, "Mr. Jack, that big steer done jumped out."

Walter said before he got to the highway he felt the truck shake a little and then he saw a big dun steer running along the side of the truck. I told Walter to go on to San Antonio and finished loading up the horses and headed home. I didn't think I could become attached to a steer but I was glad to see Sancho back with the cows and calves the next morning.

◄──────►

Dainty was a "corriente" roping steer from Mexico that I used to demonstrate the squeeze chutes I manufactured. I picked this particular steer because he was usually the first one into the roping chute and he had an impressive set of horns. The headgate on my chute didn't have a very big opening and some ranchers were afraid their horned cattle couldn't get their heads through. Dainty's horns were longer than most and he could easily get his head through the headgate. I took him to the pens at our house in town to gentle him. My daughter Kay named him "Dainty" because, as she said, he wasn't.

I put one of my squeeze chutes on a flatbed trailer and built a small pen right behind it. I loaded Dainty into the pen and went to the Uvalde County Livestock Show with my oldest son, Bob, where I had rented a space to display my livestock equipment. We backed the trailer into place and gave Dainty some hay to eat. When the first rancher came by I opened the rear gate on the chute and Bob gave Dainty a quick zap with a stock prod. Dainty walked into the chute and easily stuck his big horns and head through the headgate.

I lightly tightened the stanchions to demonstrate how well the headgate immobilized an animal then released them. Dainty backed up, pulling his horns through the opening without touching a thing, and returned to his pen. The next time Bob just walked back to the pen with the stock prod and Dainty quickly got up and walked into the chute. From that point on all I had to do was open the rear gate and Dainty would walk into the chute and stick his head through the headgate. People really got a kick out my "trained" steer.

Dainty was really gentle for a corriente steer and after a few shows he was as docile as a lamb. I took him to the San

Antonio Livestock Show where he was a big hit with the kids. One year a group of children from a special education school came to see Dainty. Most of the children were in wheelchairs and although they had seen lots of animals that day they hadn't been able to get close to any. They were so excited about being close to Dainty that I let them feed him a piece of cottonseed cake. I picked each one of the children up while Dainty ate the cake out of their hand. You should have seen the look on their faces as that big old steer licked their hands.

The next day, a second group of children from the same school came straight to my display. The children from the day before had told them about Dainty and that was the first thing they wanted to do. Once again, I let each child feed Dainty a piece of cottonseed cake. They thought it was great and Dainty was pretty happy about it too. Every year from then on children from the same school came by my display to feed Dainty. What was interesting is that every one of those kids was fascinated by his big horns and wanted to touch them. I would put Dainty in the headgate so they could feel his horns. They thought it was the highlight of the show.

Finally I cut back on going to livestock shows and retired Dainty. For years after that people asked me about Dainty, my trained corriente steer.

THE BEST HORSE
I EVER OWNED

Good horses have been important to my family for several generations. My grandfather, H.W. Kingsbery, raised quality horses in Coleman County, Texas before the turn of the century. My father, Howard Kingsbery, continued the family tradition on the Kingsbery Ranch until his three sons went into the Army during World War II. He raised and trained several top polo ponies that were used in some of the major polo matches in the United States. One horse he trained was used in an international championship match between the U.S. and England.

The best horse I ever owned or ever rode was named Bobo. He was born in 1942 and raised by Jap Holman near Sonora, Texas. Bobo was sired by his well-known Thoroughbred stud named Raffles and out of a Red Bug mare named Little Ann. Bobo was a full brother to James Hunt's good Quarter Horse stud, Little Red Raffles, a Triple A racehorse. Raffles also sired a top cutting horse named Eddie that was owned by Dub White of Mason, Texas.

Jap Holman sold Bobo as a racehorse prospect. He was taken to New Orleans and raced at the Fairground racetrack as a two-year old. Since the Fairground was a Thoroughbred track Bobo's papers had to be "doctored" to make him eligible to race there. After winning three consecutive races against some of the best two-year-olds in the country Bobo's papers were "rechecked" and he was disqualified. He returned to Texas and won several Quarter Horse races as a three-year-old in quarter-mile and three-eighths-mile races.

Bobo was sold and trained as a cutting horse, winning a novice cutting competition in Fort Worth. Then he was sold to a calf-roper and, to no one's surprise, he turned out to be a top calf-roping horse. I saw Bobo at a rodeo where I was roping. A fellow was roping calves off him and I could tell that he was outstanding. His action and movement, and the way he caught his calf and worked the rope was just about perfect. I asked a lot of questions about him and found out real quick he wasn't for sale at that time.

In 1947 Edgar Kincaid bought Bobo as a five-year-old to breed to his ranch mares at his ranch north of Sabinal, Texas. Mr. Kincaid had some good brood mares and he wanted a top stud so he paid fifteen hundred dollars for Bobo, which was a lot of money for a ranch stud back then. In 1952 the drought was in full swing and every ranch in South Texas was out of grass. One day I was visiting with Mr. Kincaid at the San Antonio Stockyards and he told me his grass was about gone and he needed to sell some horses, including Bobo.

I had seen Bobo at Mr. Kincaid's ranch and recognized him immediately. Over the years I would see him as I drove by the ranch and wished I could someday own him. I asked Mr. Kincaid how much he wanted for Bobo. He said since Bobo was ten years old he would take $200 for him. I'm sure Mr. Kincaid would have asked for more but he knew I loved good horses and would give Bobo a good home. I immediately said I would take him and wrote Mr. Kincaid a check. He told me Bobo was with the mares in the horse pasture east of the highway and that he was easy to catch. He also said Bobo hadn't had a saddle or a bridle on since he had bought him, or had his feet trimmed. I thanked Mr. Kincaid, shook hands and told him I would pick up Bobo in a few days.

I was ranching below Batesville at the time and roping calves in rodeos with two friends, Prince Wood and Bob Woodward. Our best roping horse had died recently and there was a rodeo in Utopia the following Saturday night. I called Prince and Bob and told them I had a new roping horse and to enter us in the calf roping. On Saturday morning I received several truckloads of cattle that arrived later than I expected. I got to the Kincaid Ranch late that afternoon and drove out in the pasture where Bobo and the mares were. Bobo walked right up to me and I put a halter on him and looked him over. His feet were in better shape than I thought they would be so I saddled him. It was the first saddle he'd had on in five years but it didn't seem to bother him. He just hopped right up into the trailer and we took off for Utopia, about 20 miles north, in a hurry.

The rodeo had already started when I drove up. As I was backing Bobo out of the trailer the announcer said, "Jack, you're the third roper. Hurry up." I quickly put a bridle and neck rope on Bobo, tightened the girth, got my rope and pigging string and got on him. I loped him a little way and hollered "Whoa" and stepped off. Bobo stopped perfectly and started backing up just like a roping horse is supposed to do. I got back on him and rode into the arena just as the announcer called my name as the next roper. I backed Bobo into the roping chute and nodded to open the gate.

The calf came flying out of the gate and Bobo took off like he had been roping every weekend for years. I threw my loop and pulled on the reins. Bobo came to a sliding stop as I jumped off to grab the calf. He kept the rope tight while I tied the calf and when I got back on Bobo the announcer said, "Twelve seconds flat!" It was hard to believe this was the first time Bobo had been ridden in five years. Bobo and I came in second that night and Prince and Bob both tied their calves in under fifteen seconds. Bobo worked perfectly on all three calves and I was beginning to think he just might be the best horse I ever owned.

Since Bobo was still a stud I figured I would have to keep him separated from the cow horses on the ranch, which were all geldings. But Bobo acted like any other gelding so I turned him out with the horses and he never fought with any of them. We were working lots of cattle at the time so I rode Bobo every day. I could always catch him anywhere; he would never walk off like the other horses when I came into the corral with a bridle. When friends would come to the ranch for a visit and their children would want to ride I would saddle Bobo for them. Our oldest daughter, Ann, had her first horseback ride on Bobo when she was only two months old, with me holding her of course.

Bobo was a light sorrel with a streak face and two white socks almost to his hocks. He stood 15.1 hands high and weighed about 1100 pounds with above-average conformation and the best disposition of any horse I had ever been around. Bobo was always on the proper lead and I don't ever remember him stumbling, even when we were riding through heavy brush that had been chained down. He had the easiest trot of any horse I had ever ridden. He was already trained as a cutting horse and was smooth and easy to ride when cutting cattle. He was an expert at cutting pairs from a herd, one of the few horses that would watch both the cow and the calf and keep them together.

I had about ten horses in our remuda and most mornings I could call them into the corral from the horse pasture. One morning the wind was blowing and the horses couldn't hear me so I walked to the back of the pasture to drive them to the corral. They usually went right in since I always put a little feed out for them. The horses were about half a mile from the corral. I started driving them and they were all walking along at a normal gait with Bobo in the rear. I was walking behind Bobo when I decided to jump on and ride him bareback to the corral.

Everything was going fine and half the horses were in the corral when suddenly a canvas tarp blew off some stacked hay nearby. That spooked the horses and they wheeled and ran out of the corral. But not Bobo, he did what he thought he was supposed to do—pen the horses. I grabbed Bobo's mane with both hands as he wheeled to head off the horses. It was a wild ride but Bobo and I penned the horses. It seemed like he always knew just what to do.

Bobo was the only horse I ever owned that had absolutely no faults and I am a hard critic of a horse's performance. When you run a yearling steer down a fence at full speed trying to turn it most of the time the yearling will duck behind your horse, go around him and keep on going. When a yearling tried to pull that on Bobo he could wheel so fast he would be facing the yearling and turn him. The first time I tried to turn a yearling like that on Bobo he wheeled so fast I was thrown out of the saddle. I managed to hang on to his neck and didn't hit the ground. Bobo stopped as soon as I left the saddle.

One of our neighbors had a fast, part-Thoroughbred horse, named Adam, that had outrun most of the horses in our area. He wanted to race his horse against Bobo but I kept telling him Bobo could outrun Adam so far it would embarrass him. One day I was helping him work cattle and as we were riding back to his headquarters for lunch we started through his horse pasture. I was riding Bobo and the neighbor said, "Pen Adam for me. I want to ride him this afternoon."

I told my neighbor that Bobo was so fast that I could pop Adam on the rump with my rope every step of the way to the corrals. We were on a dirt road that ran straight to the corrals, which were about a quarter mile away. My neighbor said Adam was so fast that I wouldn't get close enough to pop him. I rode up behind Adam and popped him on the rump and he took off like a shot. I spurred Bobo and he went into high gear. I popped Adam every step of the way to the corrals and just as he went into the corrals I roped him and turned the rope loose. After that my neighbor never mentioned wanting to race against Bobo.

I only owned Bobo for five months when he got sick with a lung infection. A veterinarian gave him some medicine but it didn't help. He was having difficulty breathing and getting worse by the hour and there wasn't anything I could do. I had Bobo in a small pasture next to our house, which didn't have a yard fence. As he was taking his last breath he walked around to the back door of the house and lay down and died. He had never come up to that door before and I believe he was looking for me to help him. It was a sad day on the Kingsbery Ranch.

    Mr. Kincaid kept all of Bobo's offspring and used them for many years as ranch horses. His foreman, Mr. Ware, told me they made the best cow horses he had ever ridden. Bobo was certainly the best horse I ever owned.

MAMA CATS AND BABY SQUIRRELS

My two brothers, Hank and Tom, and I were pretty typical of ranch youngsters during the 1930s. We had lots of hard work to do but also time to play and be mischievous.

Tom and I usually teamed up on most things; work, play and getting into trouble. There were always animals to ride, such as the milk pen calves after we had milked the cows or buck sheep when they were cut off from the ewes. We would run them one at a time into a chute and take turns riding them about a hundred yards into the pasture before jumping off (or falling off) and going back for another ride. We always had horses to ride so even though it was during the Depression we were seldom depressed.

When Tom and I were about ten and eleven we were riding in through the sheep looking for and doctoring screwworms. I saw a gray tree squirrel run out of a hollow in an oak tree so I rode over and looked in the hollow. There were four baby squirrels about a week old. Tom and I decided to take the baby squirrels home to see if our two mother cats would raise them.

The cats were about three years old and had been littermates. We had named them Lockstone and Schrader. I don't remember why we picked those names. They had each had several litters of kittens, always within a day or two of each other in the same nest. At the time we found the squirrels the two cats had kittens a week before in the cottonseed bin at the barn.

There were no gray squirrels in the oak trees around the barn but we did have lots of ground squirrels, which Lockstone and Schrader would catch and eat whenever they could. We didn't much believe the little squirrels would survive the adoption procedure but we wanted to try it anyway. Both cats were nursing their kittens when we introduced the baby squirrels. Tom and I sat down beside the cats, pushing their heads down whenever they raised up to see what was going on. We stayed next to the cats all afternoon trying to make sure the cats accepted the squirrels. About sundown the squirrels got hungry enough to begin nursing the cats. Tom and I took turns staying by the cats so we could eat supper and do our chores. About midnight Mother came out and made us go to bed.

Early the next morning Tom and I made a beeline for the barn. We just knew there wouldn't be anything left of the squirrels except a few hairs and bones. To our amazement the little squirrels looked contented and well fed. The mother cats had accepted them completely. There were seven kittens and four baby squirrels in all. As they got older it was really a sight to watch them all play together. The kittens could hold their own playing on the ground, but when the squirrels would climb up in the trees the kittens didn't have a chance to catch up with them.

At the time we were milking a bunch of Jersey cows and separating the cream from the milk with a hand-cranked separator that was in the back room of our house. We put the cream in the eight-gallon cans that Dad would take to the Banner Creamery in Santa Anna twice a week. Under one of the oak trees near the barn there was a flat pan that Tom and I would fill as we carried buckets of milk to the house after milking the cows. The cats and squirrels would usually be waiting for their milk handout as we came by and quickly drink it up.

One Sunday morning we were running late with the milking so Dad sent Tom and me to the house to dress for Sunday School while he finished the milking and turned out the cows. He started toward the house with the last two buckets of milk. Since he was in a hurry he passed by the cats and squirrels without pouring any milk into their pan. Tom and I happened to look out our bedroom window as Dad hurried toward the house. He was about halfway there when one of the squirrels came running after him and leaped onto the rim of the milk bucket.

The squirrel's momentum carried him over the rim and right in to the bucket with a big white splash. Dad set the buckets down, grabbed the squirrel by the back of the neck and tossed him to the ground. We could read Dad's lips and he wasn't very complimentary of the squirrel. Then he looked around to see if anyone had seen what happened, picked up the buckets and carried them into the house. Nothing was ever said about the incident.

To get cream for our cereal, coffee and for cooking Mother would always hold a pitcher under the spout of the separator to get the thickest cream. Needless to say, Tom and

I didn't eat any cereal for several days!

For years our family and friends enjoyed the antics of the mixed family of cats and squirrels. Incidentally, Lockstone and Schrader never tried to catch ground squirrels after they adopted the tree squirrels.

THE OLD BUGGY

When I was about ten years old I would often visit my grandmother, Mrs. Woodward, who lived in Santa Anna, Texas. Her neighbors, Mr. and Mrs. Culverwell, owned an old buggy they stored in their barn after buying a car. When my brother Tom and I would visit the Culverwells we would look over the buggy from one end to the other. It was in excellent condition with a complete harness lying on the seat.

Tom and I were really fascinated with the buggy and since one of our old cow horses, Rollo, had pulled a buggy in his younger days we wanted the buggy. The two of us kept trying to convince our father that we really needed that old buggy. Dad finally said we could buy the buggy if the Culverwells would sell it and we had enough money. That was in the early 1930s and times were hard. We pooled all our money, but all we could come up with was twenty-seven dollars. We just knew it wouldn't be enough because the buggy had been an expensive one and it was still in excellent condition.

We finally got up enough nerve to ask Mr. Culverwell if he would sell the buggy and, if so, how much he would take for it. Mr. Culverwell looked us over carefully and very slowly said, "Five dollars." We were surprised and thrilled. We immediately gave him the five dollars, not letting him see the other twenty-two dollars. The buggy was ours! We could hardly wait to go home and ride Rollo back to Santa Anna and get our buggy.

A few days later Mother drove Tom into town and I rode Rollo the eight miles to Santa Anna. I rode bareback and still don't know why I didn't saddle Rollo and bring the saddle back in the buggy. I finally got to the Culverwell's where Tom was waiting for me. We had never harnessed a horse to a buggy so it took us a while to get Rollo hooked up. We put the folding top up for shade and proudly drove down the main street of Santa Anna in our newly acquired method of transportation. Rollo pulled the buggy like an old hand, but when he got back to the ranch and drove around the barn, two of our young horses took a look at the new contraption and jumped over the fence in fear.

We had a great time with the buggy for several months, taking our friends for rides and even using the buggy to pen the milk cows. Rollo was pretty old and we began to realize we were working him too hard and needed a replacement. We had several good ranch horses to choose from and picked a horse named Sundown as Rollo's replacement. Sundown was about four years old and the gentlest horse we owned. Even though Tom and I were only nine and ten years old we had a lot of experience with horses. We knew we had to break Sundown to a wagon before we could hook him to a buggy.

We hitched two of our workhorses to a wagon and drove out in the middle of a big field. Tom drove the wagon while I rode another horse and led Sundown. We took the harness off one horse, put it on Sundown and hitched him to the wagon with the other horse. I rode next to Sundown on my horse with a lead rope to help hold him until he got used to pulling the wagon. We worked Sundown with the wagon for several days until we decided he was ready for the buggy.

Tom drove the buggy out in the middle of the field and I led Sundown to it. We took the harness off Rollo, put it on Sundown and hitched him to the buggy. I was on my horse and had a lead rope on Sundown to help hold him if the tried to run away. He was a little spooked at first and tried to run a few times, but with Tom in the buggy and me holding him from my horse he was easy to slow down. We worked Sundown all morning in the big field and again that afternoon until he was going well and didn't need my horse to help hold him. By late afternoon Sundown was doing so well we decided to drive him up to our house and show Mother what great horsemen her two little boys were.

We were both in the buggy as we drove right up to the side of the house and called for Mother. She heard us and went into the bedroom so she could raise the shade to see what we wanted. It was one of those spring-loaded pull-down shades. We had driven Sundown right up to the bedroom window and when Mother raised the shade it slipped out of her hand and flew to the top of the window with a big pop.

The sudden noise scared Sundown out of his wits and he took off like a shot. We managed to steer him through two gates and into a big pasture but he was running as fast has he could and even pitching some. We were steering as much as we could around the trees and bushes but he wasn't slowing down a bit. We flew down a little hill and Sundown made a quick turn to left, causing the buggy to smash into a tree. Tom and I went flying and hit the ground rolling.

It was a heck of a wreck, but fortunately Tom and I weren't hurt. The buggy was smashed to pieces, but the front wheels and shaft were still hooked up to Sundown as he kept running, headed back to the barn. Mother saw the beginning of the runaway buggy so she came running to look for us. Just as she got to the pasture she met Sundown flying toward the barn with nothing left of the buggy but the two front wheels. She was terrified until she saw us walking toward the house.

Just before Sundown reached the barn he hit a big bush and broke the shafts off, leaving the two front wheels intact. He finally stopped when got to the safety of the barn. We took the harness off and turned him back into the horse pasture; his days as a buggy horse were over. Later Tom and I build a little platform on the front wheels. Using two-by-fours for shafts, we drove Rollo many miles with that little cart. But it wasn't the same as the nice old buggy.

◀──────▶

Not too long after the buggy incident, Tom and I were riding Rollo in the lower pasture to see the new colts. Tom was in the saddle and I was riding behind, holding on to the saddle strings. We had a Thoroughbred stud named Donnie that had raced in New York for several years and later became a top polo pony. Dad bought him to breed to our ranch mares for good cowhorses and polo ponies to sell. Donnie ran with the brood mares in the lower pasture and was a gentle good-natured horse that Dad sometimes rode.

That morning Donnie, the mares and the colts were on the other side of a big tank. A creek had been dammed up to make the tank and when the tank was full it backed water up the creek all the way to the fence. The creek was about 20 feet wide and five or six feet deep. Tom and I rode around the tank and up the creek where the mares were grazing. As we were watching them they walked up to the creek and started drinking water. We thought it would be exciting to watch the colts swim across the creek, so we rode up real fast hollering loudly. The mares started crossing the creek to the other side with the colts following. It wasn't deep enough for the mares or Donnie to swim but the colts had to. I guess we enjoyed watching the colts swim because we rode back around the tank and drove the mares and colts across the creek again.

We decided to do it one more time, but this time it was a different story. As we started to drive the mares and colts across the creek for the third time Donnie decided enough was enough. He put his ears back, opened his mouth and charged. He never slowed down as he hit Rollo broadside and grabbed him by the neck. The collision knocked me off Rollo and under the two horses. They were kicking and biting each other right over the top of me. I don't know how I kept from getting stepped on or kicked. I was lucky Donnie didn't attack me. Tom managed to stay on Rollo and in a few moments Donnie wheeled and ran back to the mares.

    I got back on Rollo and we rode out of the pasture in a hurry. Rollo had some bad bites and several knots from being kicked so we had to tell Daddy what had happened. He said we were lucky not to have been hurt and warned us not to ride in the pasture unless he went with us. A stallion can be very protective of his mares and offspring.

ROPING STORIES

In the days before ropes were made of nylon and other synthetic materials, ranchers and cowboys used manila ropes made from the fiber of the abaca plant. A manila rope did the job but would break at times when roping big animals.

Since that was the only kind of rope available we had to buy them in different diameters for different jobs. For roping sheep and goats a cowboy would use a quarter-inch diameter rope. For cattle we would use three-eighths to half-inch diameter ropes. For big cattle and horses a five-eighths diameter rope was sometimes used.

The main problem with manila ropes was the variation in stiffness depending on the climate. In wet weather it would be too stiff and you wouldn't be very accurate when roping. In dry weather the rope would be too limber. This didn't make a lot of difference when a cowboy was just working on the ranch or in the corrals where a miss or two didn't matter. The big problem was when you were roping in a rodeo. You needed to have the rope feel just right as you backed your horse into the roping chute. Timed rodeo events were so competitive that everything had to be just right to win.

We used a lot of different ways to make our ropes feel right when our time came to rope. Some ropers had a canvas rope bag that did a pretty good job of keeping the ropes limber. Others used galvanized sheet metal canisters with a tight-fitting lid that worked pretty well. The best rope canisters had a cup in the middle so you could put a damp cloth in with the rope if the humidity was real low. The moisture from the damp cloth would keep the rope from getting too limber. On wet days we would tie our rope under the hood of our pickup or car as we headed to the roping. The heat from the engine would dry them by the time we got to the rodeo so the rope would be just right for roping.

Manila ropes were sold in big coils that you would cut to length. After you tied a loop in one end and a horn loop in the other end you had to stretch them overnight to get the coils and kinks out. There were lots of ways to stretch a rope. Some cowboys tied one end to a tree or post and used a wire stretcher tied to another tree or post. Others would tie each end to a tree or posts and put a heavy weight in the middle. Another way to stretch a rope was to tie one end to a tree and the other end to the trailer hitch on a vehicle and pull it tight, then lock the brakes and leave it stretched overnight.

That method got me in trouble once. I was attending Texas A&M College in 1948 and my good friend, Prince Wood, had just bought a new rope and needed it stretched. There was a tree close to our dormitory so he tied one end of the rope to it and I backed my car near the tree. Prince slipped the loop over my trailer hitch and I drove forward until the rope was tight and locked the brakes. The next morning I went out to the car and there was a ticket on the windshield. According to the ticket the violation was for not having brakes on my car.

I went to the campus police station to ask about the ticket. The campus cop who wrote the ticket was in the office and he told me it was dangerous to have a car with no brakes. That was the only reason he could think of why my car was tied to a tree! The police chief, who knew me, was also in the office. When he heard his officer tell me the reason for the ticket he looked real embarrassed. He told me to give him the ticket and he would take care of it. I told him that I would keep the ticket; the campus newspaper—The Battalion—might be interested in it. As I walked out the door the chief said, "Jack I sure wish you would give me that ticket." I took the ticket to The Battalion office and they ran a story on the "car with no brakes." The Houston Post and the Fort Worth Star-Telegram picked up the story and also ran it.

Not too long after that nylon ropes came into being and we didn't have to worry about diameters or humidity or traffic tickets. Writing this story reminded me of my old rope canister. I thought it was still in the storeroom at the barn so I went out to look for it. There it was, on the back of a shelf covered with dust. It hadn't been opened in about thirty years so I dusted it off, opened the lid and there was my manila calf rope, as good as new. I had roped lots of calves with it and it brought back a lot of roping story memories.

◄─────────►

I began roping when I was four years old, starting with chickens and milk pen calves. In my teens I roped in the boys' breakaway calf roping at the annual rodeos in Santa Anna and Coleman, Texas. At Texas A&M I was on the rodeo team and was the champion calf roper at the A&M rodeo in 1948. As I grew older I got slower and the young ropers got faster, so I finally gave up rodeo roping in the 1970s when I was in my fifties.

I still love to rope and over the years have caught a javelina, a coyote and, once, a big buck. It happened in 1949 when several of us were roping out some outlaw cows on the Valdina Ranch north of Sabinal, Texas. The ranch had a high fence around it and part of the ranch didn't have a lot of brush. I was on a fast horse and we were chasing an outlaw cow that was headed for the high fence. Just before she got to the fence a big buck ran out of an oak mott and headed down the fence. A deer can outrun the fastest cowhorse but since we were going full speed and had an angle on the deer I decided to rope him.

As the buck went by me I threw a long loop and caught him by one horn. A big buck can be dangerous when cornered and I was lucky that one of the other cowboys came up and roped the buck around the horns. The buck fell down and I jumped off my horse, held him down and took both ropes off his horns. He jumped up and ran off, no worse off from the experience. I roped that deer on a sudden impulse and could have gotten hurt, but at least I can say I roped a deer. It didn't take my cowdog, Mitzi, long to find the outlaw cow again so we could rope her and haul her to the corral.

Later on, while ranching west of Crystal City, I was rounding up cattle and Leroy Fatheree was helping me. We had to ride across a big field to get to the pasture where the cattle were. Suddenly a coyote jumped up right in front of us. Since both of us were riding fast horses we took off after the coyote to rope him. A horse can outrun a coyote but a coyote can duck and dodge, making it almost impossible to rope one. Leroy and I kept the coyote circling in the field so it wouldn't have a chance to duck under the fence and get away. After about the third circle around the field I got close enough to throw my loop and caught the coyote around the neck. Leroy got off his horse and turned the coyote loose. I added the coyote to my list of animals I had roped.

Another time I was checking cattle alone and rode up on several grown javelinas in some fairly open country. They headed toward a brushy creek about a quarter of a mile away and I decided to rope one. I was riding my good quarter horse, Bandy King, and he caught up to the javelinas in a few jumps. I picked out the largest one and roped him. He was a big boar and my loop went over his neck and one front leg.

When I took off after the javelinas it didn't dawn on me that if I roped one I would have to get the rope off. Javelinas can be dangerous when cornered and mine was really popping his teeth and making runs at us. Bandy King was so quick he could get out of the way but I'm sure he wished I would stick to roping cattle. I kept popping my loop hoping it would come off the javelina's head. Since the loop was behind the front leg it wouldn't come off.

I then rode around a big mesquite tree a couple of times and pulled the javelina up to the trunk of the tree. The tree had limbs and one almost touched the ground. I got off Bandy King and climbed up the limb to the trunk of the tree. When I was about four feet above the javelina I broke off a limb and used it to push the loop off the javelina's head and away he went.

I guess I learned something from that incident. Recently I was riding by myself when a big feral hog jumped up right in front of me and headed out across some open country. Instinctively I grabbed my rope and took off after him. Then common sense took over and I pulled my horse up and put my rope back on the saddle. There I was by myself, miles from anybody, seventy-five years old and riding a fourteen-year-old horse. It was really tempting to rope the hog, but I just sat there on my horse and watched the big hog run across the pasture.

MAD COWS & BAD BULLS

Doctoring screwworms was an everyday job for a rancher in South Texas during the 1960s. One day our second oldest daughter, Kay, was riding with me checking the cattle for what we called "wormies." She was 13 years old and riding one of our good cowhorses, Mr. Pleasure Bar, sired by my AAA Quarter Horse stallion, Pleasure Bar. I was riding a gray gelding named Sundance that was sired by another one of my top cowhorses, Bandy King.

Kay and I found a cow with a three-day-old calf that had screwworms in its navel. In most cases I would rope the calf and the other rider would keep the cow away while I got off my horse to doctor the calf. A good mother cow will just about kill anyone who messes with her baby. This mama cow had been dehorned but still had a short nubbing horn left that could do some damage. As soon as I roped the calf it bawled and the mama cow came right up to us and hooked my horse. Kay would try to drive her away but she would run at Kay and try to hook her and her horse. Kay would have to spur Mr. Pleasure Bar to keep from getting hooked.

I quickly realized that I couldn't get off and doctor the calf in the pasture. Kay was too young to rope the cow and hold her while I did the doctoring. We were about three-quarters of a mile from the corrals and I needed to get the cow and calf to the corrals so I could separate them. We were right by a galvanized gate in a partition fence so we drove the cow through the gate and Kay held the gate shut from her horse. I pulled the little calf up, put it across the front of my saddle and headed for the corrals. As I rode off the cow butted open the gate and took off after the calf and me. Sundance was going at a fast lope but the cow was right beside me hooking at my horse and me. She wanted her calf back.

Kay caught up on her horse and tried to keep the cow away, but she charged them instead. As Kay wheeled Mr. Pleasure Bar the cow stuck her head between the horse's hind legs and almost flipped him over, throwing Kay out of the saddle and onto Mr. Pleasure Bar's neck. Kay grabbed his neck and hung on for dear life. The cow wheeled and came back after Sundance and me. Kay managed to get back into her saddle without falling off the horse.

The cow stayed right on Sundance's tail all the way to the corrals, trying to hook me. She bruised my leg in two places with that nubbin horn of hers. Sundance kicked at her several times but that only made her madder. The corral gate was open so I rode right in and lowered the calf to the ground. My rope was still around the calf's neck so I rode out of the corral real fast and Kay shut the gate. Using the rope I pulled the calf up to the gate, opened it quickly, grabbed the calf and latched the gate, leaving the mad mama cow inside the corral. As Kay and I were doctoring the calf the cow kept hooking the gate, trying to get to her calf and us. I figured with a mama like that, the calf was safe from coyotes and other predators.

Another time I was riding my good cowhorse, Bandy King, looking for wormies on the Von Rosenburg Ranch northwest of Crystal City. I found a cow with a week-old calf that had screwworms in its navel. The mama cow was a big black and white Brahman-Holstein cross with tall curving horns. I roped the calf and could tell real quickly that the cow wasn't going to let me get down and doctor it. I pulled the calf over to a big mesquite tree with branches running out sideways about head high as I sat on Bandy King. I tied the end of my rope to the tree branch and rode back to my pickup and trailer.

I tied Bandy King to a tree and drove the pickup with the trailer back to the tree where I left the calf. I drove under the limb where the rope was tied, stopped the pickup and got into the bed of the truck with my worm medicine. As I started pulling the calf into the back of the truck the mama cow was hooking at me over the side of the pickup. When I pulled her calf into the bed of the pickup the cow reared up, put both front legs over the side and tried to jump into the back of the pickup with me. I jumped out the other side of the truck as she tried to jump in again. She couldn't quite get over the side of the truck so she backed up a little way to get a running start. I quickly climbed back in the bed, lifted the calf out of the pickup and put it back on the ground. This wasn't going to work.

I pulled the pickup forward until the back of the trailer was under the limb where the calf was tied. The trailer had an open top so I climbed into the trailer and untied the rope from the limb. I opened the trailer's gate just wide enough to pull the calf inside and shut the gate. As I was doctoring the calf the mama cow kept hooking the sides of the trailer trying to get to me. When I finished doctoring I let the calf out of the trailer and they headed off into the brush. I guess lots of people admire how well mama cows protect their babies. At times I wish they weren't so motherly.

Another time I was checking cattle while riding a three-year-old horse I had raised named Fireball. I had only been riding Fireball about eight months so I was using a hackamore instead of a bridle and bit. The ranch was pretty brushy with four creeks running through it. The creeks all had water in them from rain a few days earlier.

I rode up on a big Angus bull and saw he had a bad case of screwworms between his upper front leg and brisket. As usual I was carrying screwworm medicine in my homemade boot-top pouch and had several options for doctoring the bull. There was a set of corrals about three-quarters of a mile away but the chute wasn't in very good shape. Another option was to drive him to the Crawford Farm less than half a mile away where there was a set of corrals with a good chute. The drawback was I would have to drive him across all four creeks to get him there.

The third option would be to rope him and tie the rope to a tree. Then I would heel him with another rope (I always carried two ropes for just such situations) and stretch him on the ground for doctoring. That would normally be the easiest way but there were two drawbacks to that option. First, I was riding a young, inexperienced horse that only had a few calves roped off him. Second, I had only bought the bull four months before and I didn't know what he would do when he was roped.

I decided to take the safer option and drive him to the Crawford Farm. It was just a little way to the first creek and the bull didn't want to cross the water. He ran up and down the bank several times before we finally drove him across the creek. The second creek was even harder because the brush was so thick. It was a hot summer day and the bull kept stopping in the shade of trees surrounded by brush and wouldn't move. I would break off dead tree limbs and throw at him to make him go in the right direction. I was having so much trouble getting him to go that I decided to go ahead and rope him. I built a loop, but the bull was so big I couldn't get up nerve enough to rope him. I coiled my rope, put it back on the saddle and tried to get him across the second creek.

After a lot of stick throwing and running we got him across the second and third creeks. The fourth creek was close to the Crawford Farm but by then the bull was getting real hot and mad. He started fighting Fireball and wouldn't cross the last creek. Once again I decided to rope him, but when I took another look at him I realized it would be foolish to rope a big mad bull on such a young horse. I decided to make one more attempt to drive him across the creek and that effort was successful.

Mr. Crawford saw me coming and opened the gate to his corral and the bull went right on in. To do a good job of treating the screwworms I needed to throw the bull down. I decided the best way to throw him would be to put him in the running chute and put a rope around his neck. Then let him out of the chute and tie the end of the rope to the mesquite tree in the middle of the corral. Then I could heel the bull from Fireball and stretch him out to doctor him properly.

It seemed like a good plan. The bull went right in the chute and I put a pole behind him so he couldn't back up. I took one of my ropes and put it around the bull's neck. Just as the loop touched his neck the bull went crazy—I don't know a better word to use. He started bellowing and jumping and kicking as hard as he could. Then he started butting the headgate and the sides of the chute. Boards started flying in all directions. Mr. Crawford had been standing close to the corrals but when I glanced over my shoulder I saw him running into his house. The bull never quit fighting until he had torn the chute completely apart. Then he started running around the corral bellowing and pitching and trying to hook Fireball.

I grabbed the end of the rope that was still around his neck and made a quick tie around the tree. The bull charged me and I barely made it over the fence. I waited until the bull was on the other side of the corral before making a mad dash to get on Fireball. I took my other rope, heeled the bull and stretched him out on the ground. With the bull unable to move I pulled up his front leg and poured the medicine on the screwworms. I took the rope off the bull's neck and, with Fireball still holding the heel rope tight, I opened the gate into the pasture. I got back on Fireball and slacked the heel rope to let the bull get up. He made a beeline for the pasture.

When Mr. Crawford saw the bull had gone back in the pasture he came out of his house laughing. He said he'd never seen a bull that wild. I apologized for the damage to his chute and told him I would repair it in a day or two. I often wonder what would have happened if I had gone ahead and roped that big bull out in the pasture on that young horse.

One day Evelyn and I were checking cattle on our ranch west of Crystal City when we found another Angus bull with screwworms in his sheath. The surest way to treat screwworms in that situation is to pour screwworm medicine down the sheath. This was a big fat Angus bull that I had recently bought and he seemed pretty gentle so we decided to drive him to the corrals about half a mile away and treat him in the chute.

We were riding two of our top cow horses and started driving him toward the corrals. As we got close to the corrals the bull made two attempts to run back into the brush. We managed to turn him both times and head him back toward the corrals. Just as he started going through the gate he wheeled, hooked at Evelyn's horse and took off for the pasture as fast as he could run.

I saw the bull couldn't be penned so I took out my rope and took off after him. I was riding Old Sorrel, a big stout horse, and just as I roped the bull he ducked to the right of a big mesquite tree trunk. It was about two feet in diameter and had been dead for years. All the limbs had fallen off but it was still ten feet tall and looked pretty sturdy. We were going too fast for me to pull Old Sorrel to the same side of the tree as the bull. With my rope tied to the saddle horn and the tree between the bull and me, I was sure there was going to be a disastrous wreck. This was one time I wished I was a "dally" roper—someone who doesn't tie his rope fast to the saddle horn.

We were side by side as we passed the mesquite tree and I was getting ready for a heck of a wreck. I was sure Old Sorrel and the bull would both be jerked down when the strong nylon rope hit the tree between us. I jerked my foot out of the right stirrup and threw it over my horse so both my legs were on the left side of the horse away from the rope. I held on to the saddlehorn for balance so I could push away from my horse when he fell. Evelyn was looking on in horror, expecting to witness a major catastrophe.

At the moment the wreck should have happened, I felt a big jolt and the rope slacked as the big mesquite tree was pulled down. I put my leg back over the horse and into the stirrup and pulled Old Sorrel and the bull to a stop. I realized we would have to heel and throw the bull right there in the pasture to treat him. Evelyn was an expert at working cattle but roping a mad bull's heels was a little above her ability. To heel him we would have to switch horses. That wasn't going to be easy because the bull wasn't as gentle as I first thought. He had charged my horse several times and was pawing the ground daring either of us to get off our horses.

Evelyn rode her horse over to mine and we made the change in a split second as the bull charged. Evelyn reined Old Sorrel out of the way and wheeled him to face the bull as it hit the end of the rope. The bull would have jerked Old Sorrel down if he had been turned sideways. I heeled the bull and we stretched him out on the ground while I doctored his sheath. I got back on Evelyn's horse and kept the heel rope tight while she rode my horse toward the bull to let slack off the rope. She got down, pulled the loop off the bull's neck and got back on Old Sorrel while I rode up to slack off the heel rope. The bull got up and trotted off through the brush.

It was just another routine ranch job with a little added excitement.

A TRUE TWISTER TALE

The movie "Twister" reminded me of an incident that happened at our ranch in 1927. Coleman County, Texas is in "Tornado Alley" and just about every ranch family had a storm cellar close to their house. They were big holes in the ground covered with heavy timber and made waterproof. When a tornado was sighted everyone would get into the storm cellar for protection.

Late one evening a big storm was headed toward the ranch so Mother took me and my two brothers, Hank and Tom, down into the storm cellar. Dad was out riding in the pasture and didn't get back to the barn until after we had gone into the cellar. He wasn't afraid of storms so he milked the cow and carried the milk to the house. We had gas lighting in the house and Dad lit the kitchen light when he went into the house.

Just then a tornado hit the house and blew the entire west wall of our house out. The kitchen was in the west end of the house and Dad ran out where the wall had been. The tornado picked him up and blew him out into the pasture. Mother had forgotten to latch the storm cellar door and it blew open. She ran up the steps calling for Dad with us three kids right behind her.

I was only five years old but I will never forget the sight of our house with an entire wall missing. The light in the kitchen was still on and we could see the kitchen stove, cabinets and table. Nothing had moved. The wall had landed a only few feet from the storm cellar door. We called for Dad and heard him answer from down in the pasture. He had been blown a good ways before grabbing a small mesquite tree to keep from being blown even farther. Dad was skinned up some, but not hurt too badly.

As he came down into the storm cellar, making sure to latch the door, he told us that the barns and sheds were still standing. Only the wall was blown out of the house and he told Mother it wouldn't be much of a job to fix the house. I remember we kept asking Dad if he had rolled along the ground like a milk can or if he was on his feet. Unfortunately I don't remember what he told us.

We kept hearing the wind howl and the rain pouring down. After about an hour the noise stopped and Dad opened the door to a brightly shining moon. We all went up the steps to look at our house. It was flat on the ground. The tornado had circled and hit our house again. The barns and sheds weren't damaged.

We had a collie named Prince that our uncle, Jack Woodward, had given us. Prince was about three years old when the tornado hit and my brothers and I were more worried about him than about our house blowing down. We called and called but no Prince. We just knew the house had crashed down on him. Dad cut off the gas to the house at the meter and we drove into Santa Anna to spend the night with Grandmother Woodward. When we got to the Chambers place, about three miles south of Santa Anna, Sweety Creek was running across the road from all the rain. We had to wait until it got low enough to cross.

The next morning Dad left early for the ranch to check on everything. The first thing he saw was Prince lying by the wrecked house like he was dead. Dad called and Prince slowly raised his head. He was all skinned up and so stiff he could hardly walk. The tornado had probably picked him up like Dad and blown him a long way. When Dad got back from the ranch that evening, the first thing we asked him was "Did you find Prince?" We were really thrilled when Dad said Prince was alive. It was about a month before Prince was back to normal.

The only contents of the house that weren't destroyed was Dad's rocking chair and my little oak rocking chair that Uncle Woodward gave me when I was two years old. We still have that little rocking chair. Dad built another house a few feet north of the one that had blown down. We continued going to the storm cellar when a bad storm came but Dad would seldom go. We had a bed in the cellar and we kids would usually be asleep when the storm passed. Mother would leave us there, with Prince, until morning.

From the day of the tornado, Prince was scared to death of storms. We only went to the storm cellar when there was a bad storm cloud coming up but Prince was terrified from the first clap of thunder. He was the first one down the steps when we headed for the cellar. Mother and Dad never allowed dogs in the house but sometimes they made an exception for Prince. When he heard thunder Prince would come to the back porch and whine. Sometimes one of us kids would run to the storm cellar and open it for Prince. If there was a bad thunderstorm with lots of lightning Mother would let Prince into the utility room until the storm was over. As soon as he was in the cellar or the house Prince would quit whining.

I remember a Mr. Carter, who came from Mississippi to run a pump station on our ranch. He made fun of the local people for being scared of storms and going into a storm cellar. Late one Saturday afternoon a tornado hit the town of Clyde, about 45 miles north of Santa Anna. It blew down a lot of houses and businesses and killed a number of people. Mr. Carter and his wife drove to Clyde the next morning to see the damage. He got back to Santa Anna that afternoon and, even though it was Sunday, hired several men to go out to the oil lease where he lived and start digging a hole for a storm cellar! The men had the hole dug by dark and put a top on it the next day.

Our storm cellar is still there at the ranch but people don't go into them like they used to. Now they watch the direction of a tornado on television. If it is heading their way then they run to the cellar.

THE WILD BUNCH

Tink Churchill, a local cow trader, wanted to sell me a small herd of cows. They were on a ranch about two miles south of Crystal City near Turkey Creek. I wasn't really in the market for more cattle but Tink kept coming down on the price until I agreed to go look at the cows.

The pasture was pretty brushy but we were able to drive around and see all the cattle. There were 27 Brahman cows, four and five years old, bred to Hereford bulls. Most of them had newborn calves and the rest were due to calve in a week or two. They were good quality cattle and the price was right, but I didn't act too interested until Tink said I could leave them in the pasture for three months. Three months of free pasturage made it a good deal so I bought the cattle.

A week later I went back to the pasture on horseback to check on the cattle. That's when I found out why the price was so good. At the first sight of me on horseback the cattle would run off like deer. I rode for several hours and never could get close enough to check any of them. Since they came from the Victoria, Texas area, which is very brushy, I thought maybe they had been worked by cowdogs. The next day I brought my two good Catahoula "leopard" cowdogs, Mitzi and Lady, with me. The dogs would try to hold them until I got close enough but the cows would scatter like quail. One of my dogs even caught and held a calf but its mother was so wild she wouldn't even come back to fight!

I was beginning to get the idea why Tink wanted to sell them to me for range delivery and was so generous with pasturage. He had some of the best cowboys in the country working for him and they handled lots of wild cattle. If Tink's men couldn't round up these cattle I realized I had a real problem. I figured I had three months to get them tame enough to pen, so I started putting out cottonseed cake for them. At first only one or two cows would come to my pickup for cake. Then we had our first freeze, which browned the grass, and more of the cows came to the cake.

Before the three months were over I had the whole herd following my pickup looking for cake. All the cows had calves by then and were doing very good. The market was up and I had a good profit in the cattle, if I could get them in a pen to sell them.

There was an old set of corrals on the ranch but it wasn't in very good shape so I decided to drive them down the highway about half a mile to a good set of corrals at Dr. Don Smith's farm. I knew the cattle would follow my pickup out on the highway and that I could shut the gate to the pasture behind them. After that I wasn't sure what would happen.

I figured I would need lots of help with this wild bunch so I hired six good cowboys and had them hide on horseback in the brush next to a highway bridge over Turkey Creek. I opened the gates leading to the Smith Farm corrals and then drove to the pasture with the cattle. I honked to call them and before long all the cows and calves were following my pickup. I drove out onto the highway right of way where I scattered cottonseed cake and then shut the pasture gate. I drove down the right-of-way past where the cowboys were hidden and got on my horse. We all started moving toward the unsuspecting cattle, seven abreast. We got pretty close before the cattle looked up from their cake and saw us.

You would have thought a bomb went off in the middle of the bunch. When they saw seven cowboys coming they took off at full speed in whatever direction they were facing. Most of them raced right past us and jumped the fence back into the pasture. Some of them ran into the brush at a nearby gravel pit and lay down, trying to hide from us. One cow left her calf and ran along the highway for almost a mile before lying down to hide in some brush next to the railroad tracks. In all my years of working wild South Texas cattle I have never seen a bunch of cattle as crazy as these. There was just no way they were going to be corralled by horseback. The only thing left to do was get them all back in the pasture.

We rode around the ones in the gravel pit and they jumped up and leaped the fence back into the pasture. One of the cowboys, Freddy Pond, and I loaded our horses in my trailer and drove near the cow that had run down the highway. We got on our horses and I roped her so we could load her in the trailer and take her back to the pasture. I paid the cowboys, but told them they should pay me for all the excitement.

I kept caking the cows and got them back to following my pickup after a few days. I did a little work on the old corrals that were on the ranch. I fixed the loading chute and wired some old aluminum irrigation pipes along the top of the corrals to discourage the cows from jumping out. A Brahman cow can clear a five-foot fence flat-footed if she wants to. Then I started putting out cake in the corrals. It took several days before all the cattle would come into the corrals. I was ready to try to ship them to market one more time.

I had one of Gene Allen's cattle trucks wait on the highway while I called them into the corrals. There was a hard wind blowing that morning and all but four cows and their calves came into the corral. Those four were too far away to hear the pickup horn. I closed the corral gate and drove out to the highway and told the driver, Walter Nation, that I was ready to load. He drove into the pasture and backed his truck up to the remodeled loading chute. Walter and I didn't have too much trouble loading the cattle. One cow jumped out of the crowding pen into another pen but we got her back into the crowding pen and into the truck. When they were all loaded Walter took off for the San Antonio stockyards and I was happy to see that bunch of wild cows gone. Cattle prices were still good and they brought a good price.

The next week I tried to call the last four cows into the corral but they knew something was wrong. They would come within a hundred yards but no closer. I ended up roping them out one at a time and sold them at the auction in Uvalde, Texas. The cows ended up making me good money but it was an experience I could have done without. I'll never forget how the wild bunch exploded when they saw seven cowboys coming at them.

THE BLIND COWBOY

I have always had good cow horses and Catahoula "leopard" cowdogs to find wild cattle in the heavy brush of South Texas. A lot of times I helped my neighbors catch wild cattle that couldn't be gathered in the regular roundups. It is hard for people who aren't familiar with the Brush Country to realize how hard it is to catch cattle if they decide they don't want to be caught.

The early ranchers in South Texas ran mostly steers instead of cows and calves. Until the 1960s nearly all newborn calves got screwworms in their navels and would die if not promptly found and doctored. The screwworm fly, also called blowfly, would lay its eggs on any wound if there was blood present. A newborn calf's navel was a sure target for the screwworm fly. The government finally eradicated screwworms and opened up South Texas to cow-calf ranching, increasing wildlife populations at the same time.

Ranchers would gather their steers two or three times a year and there would always be a few wild cattle that avoided the roundup. These cattle were called remnants and if not caught and hauled out would escape the next roundup, taking a few more cattle with them. It was really a chore to catch the remnant cattle and some cowboys had more ability for chasing cows through the heavy thorn-covered brush than others. We wore heavy denim brush jackets, thick leather leggings (never called "chaps") and had heavy toe fenders (also called "tapaderas") on our stirrups to keep the brush from tearing through our boots. We wore leather chinstraps on our hats to keep from losing them during the chase. It was almost like donning armor before a battle.

The brush and trees we had to ride through were mesquite, huisache, catclaw, blackbrush, whitebrush, retama, granajeno, and guajillo. If that wasn't bad enough there was plenty of big prickly pear and tasajillo bushes. Except for the whitebrush and guajillo, everything was covered with sharp tough thorns.

I had helped rope out a lot of remnant cattle over the years, some of my own and some for neighboring ranchers. We would usually use our cowdogs to find the cattle and then we would take off after them through the brush and try to rope them. The animal would be led out, usually stubbornly resisting all the way, to an area where they could be loaded into a trailer. Since we would rope out several wild cattle before getting the trailer we would have to tie each one so it couldn't get away. We would either "neck" them to a tree or "sideline" them.

One cowboy alone can neck a cow to a tree but it's a lot easier if you have a partner to help. Both of us would rope the animal and pull it up to a tree, one on each side, until its head was against the tree. We would then tie the animal to the tree with a neck rope that was doubled in a way that will keep it tight without choking the animal. We would wrap the ends of the neck rope to the tree and tie them in a knot that would let the animal move around the tree without choking.

To sideline an animal you tie a front foot to a hind foot on the opposite side about a foot apart. This way the animal can stand up but can't go very far, mostly moving in a circle. It will still be about where you tied it when you come back to haul it out. Sidelining requires you to throw the animal, either by heading and heeling it if you have a partner or tripping it if you are by yourself.

Tripping an animal is done by roping it around its neck then running past the animal, throwing the rope over the animal and behind its hind legs and jerking it down hard. You jump off your horse, hollering at your horse to keep going so the animal can't get up. You quickly grab the animal's front leg to hold it down and holler "whoa" so your horse will stop, keeping the rope tight. You can then tie the feet together and start looking for another one. A good cow horse is a definite necessity for one-on-one sidelining.

To haul the wild cattle out you back the trailer close to the animal, open the tailgate and put a rope through the front of the trailer and around the neck or horns of the animals. Then you untie the animal from the tree or take off your sideline rope and, using your horse, pull the animal into the trailer. Sometimes it takes a whole day just to find, catch and haul out two or three wild ones. Pastures on South Texas ranches can be several thousand acres in size.

◄─────────►

One of my ranching neighbors was Jay Whitecotton, a good friend with a booming voice, a reckless spirit and a passion for flying. He was running cows and calves as well as steers on his ranch. Jay asked me help catch three wild Brahman cows that his cowboys had been trying to pen for two years without success. Jay told me that I didn't need to bring my leopard cowdogs even though the cows were in a two-section (1280 acres) pasture in the brushiest part of his ranch. He said he would locate the cows with his airplane and lead me right to them. Jay had owned his little single-engine plane for two years and really loved to fly.

   I took two of my best cow horses because sometimes a horse gets briefly crippled running through the brush. I visited with Jay for a few minutes at his ranch headquarters where he had a landing strip. He told me how to get to the pasture and then he got in his plane. By the time I got to the pasture where the cows were hiding Jay was flying in a circle over the middle of the pasture. That's where the cows were. I had my loop ready when I rode into the thicket that Jay was circling in his plane. I rode almost on top of three big white Brahman cows that were lying down hiding. Once they realized I saw them they took off just like deer.

   One split off and I ran her about a quarter of a mile and roped her, threw her down and sidelined her. By the time I got back on my horse and rolled up my rope Jay was circling over another part of the pasture. I rode up within 20 feet of the two remaining cows, which were lying down trying to hide from me, when they both took off. I ran them about the same distance as the first cow before roping one and sidelining it.

   I was ready to catch the third cow, thinking that using a plane to catch wild cows wasn't such a bad idea. This was long before helicopters were routinely used to round up cattle so Jay and I were way ahead of the industry. By then Jay was circling above a heavy stand of brush that was as thick as a jungle. I rode in knowing the cow was lying down in the thickest part of the brush, but I couldn't see her. As I was riding back and forth looking for her Jay flew down low, cut his engine off, opened the window and hollered out in that booming voice, "Are you blind? She's right there close. You almost ran over her!" Then he started the engine and gained altitude.

   It was a lot easier to see the white cow in the thick brush from above than from horseback. The brush was taller than my horse and me and almost impenetrable. I kept looking, knowing the cow was almost close enough to touch, for another five minutes. Jay flew down low again, cut his engine and yelled out loud and clear, "You must be blind! You need my glasses.

She's right down there!"

About that time I came so close to the cow that she jumped up and took off. I roped and sidelined her and headed back to the pickup. By the time I got there Jay was already there in his pickup so we loaded the three cows and hauled them to his corral. He really kidded me about not being able to see that last cow. He said I got within fifteen or twenty feet of her several times.

WHO GOT RATTLED?

I have hunted a lot of deer in my life and have taken a lot of people deer hunting. It can be interesting to watch other hunters' reactions when they are about to shoot a deer.

I was ranching near Batesville when a friend from Uvalde asked if I would take a friend of his hunting. I said I would and the next day the man came to the ranch. He was a top executive with the Phillips Petroleum Company from Bartlesville, Oklahoma. He was excited about hunting and told me he had killed his first deer the year before from a stand in Oklahoma. While we were talking about deer hunting I told him about rattling up bucks. During the rutting season bucks will come to the sound of someone hitting and rattling two deer horns together. If done right it sounds like two bucks fighting over a doe. Another buck thinks he can sneak up and steal the doe while the other bucks are fighting.

The Oklahoman really wanted me to rattle up a buck for him. The next morning was perfect for rattling—cold and clear with no wind so the sound of horns would carry a long way. I put the Oklahoman in a big mesquite tree facing the direction I thought a buck would come to horns. He was sitting about eight feet above the ground and had a lever-action 30-30 Winchester rifle that he told me had belonged to his father. I told him to be on the lookout as soon as I started rattling because a buck might come sneaking in slowly or running at full speed. As I was explaining all this he was really getting excited and could hardly wait until I started rattling.

I hid in some bushes a little way from the tree and started hitting and rattling the deer horns together. Sure enough, after a few minutes, I could tell by the look on his face that he had seen a buck. From where I was I couldn't see the deer but knew he was looking at one. All a sudden he started pumping the lever on his rifle real slowly. Every time he pumped the lever a shell ejected from the chamber and hit the ground below him. I couldn't believe my eyes, but I couldn't say anything because the deer would hear me and run off. He pumped the lever until all six shells had fallen out and then a couple more times for good measure. Then he raised the 30-30 to his shoulder, aimed carefully and pulled the trigger.

Of course the only thing that happened was the sound of the hammer hitting the firming pin with a loud snap. At the sound I saw a big buck wheel and take off at full speed. The man grunted in surprise, pumped the lever, pointed it at the running deer and pulled the trigger. Again the hammer snapped on an empty chamber. After a few minutes the man climbed down and said in disgust, "I forgot to load my gun." I said, "No sir, you just unloaded it. Look on the ground."

He looked down and saw the six shells on the ground. He couldn't believe that he had gotten so excited that he would work the lever and eject all the shells out of the gun without being aware of it. He was able to run a big corporation but completely lost control when he saw that big buck. I assured him I wouldn't tell anybody and I never did until now—forty-seven years later.

◄——————►

During the hunting season a lot of deer are wounded by hunters who don't "forget" to load their guns. The best way to find a wounded deer is using a dog to trail it. One of my Catahoula (also called "leopard") cowdogs, named Mitzi, was a top deer dog. I would put her on the trail of a wounded deer and she would trail it until she flushed it. Then I would turn my other two other cowdogs, Lady and Blue, loose to help. Blue was a good catch dog and he would grab and hold a wounded deer. A deer that isn't wounded very badly will run for a long ways and having a dog that will grab and hold the deer saves a lot of time and running. Lady was a daughter of Mitzi and could also help trail and hold a deer.

One time I got a call from the Holdsworth Ranch near Batesville to bring my cowdogs and trail a wounded deer. When I got to the brush-covered pasture where the deer had been shot the ranch foreman was there with several hunters from Houston. The man who had wounded the buck was a business executive and was really excited about shooting such a big buck. He said it had horns as big as an elk and he sure hoped the dogs could find it.

There were a few drops of blood on the ground so I let Mitzi get a scent and then turned her loose. She took off and after about ten minutes we heard her barking so I turned Lady and Blue loose to help Mitzi. I could tell from the way the dogs were barking that the deer was moving real fast so we started following on foot as fast as we could through the thick brush. In a little while the dogs changed from a trail bark to baying which meant they had stopped the deer. As we got closer to the dogs and deer the brush was so thick you couldn't see ten steps in any direction.

When I guessed we were about seventy-five yards from where the buck had stopped I told everyone to wait and I would work my way into the heavy brush and shoot the buck. The man who had shot the deer told me he wanted to finish the job. He said it wouldn't mean anything to him if he didn't take the final shot. I told him it was really dangerous to shoot the buck with my dogs so close. Blue was probably hanging on to the deer by then and I didn't want any of my dogs to get shot. The hunter insisted that he would be very careful and wouldn't shoot until I told him to.

He didn't seem too excited so I reluctantly agreed to let him shoot the buck. He and I eased into the heavy brush toward the sound of the dogs. I could tell they were really having a battle. The hunter had an automatic 30-06 rifle and I cautioned him again not to shoot until I said to. As we got closer I could see the dogs and the deer in a small opening in the heavy brush. Blue had the deer by the neck, Lady had a hold of a hind leg and Mitzi was darting in and out. They were going round and round locked in combat. It was a huge buck and I could see why the hunter wanted to finish it off.

As we stepped out of the brush into the opening the hunter suddenly threw his rifle to his shoulder and started shooting at the deer and the dogs. After the first two or three shots I reached out and jerked his rifle up just as another shot went off, this time toward the sky. I took the rifle away from him and waited until the dogs were clear. Then I killed the deer with a shot to the neck. Fortunately none of the dogs were hurt.

The hunter's face was white as a sheet and he just sat down on the ground. He was so shocked by what he had done that he couldn't say anything at first. Then he apologized and said he had just lost control when he saw the deer. I told him I wouldn't tell the others what had happened and he managed to stand up by the time they arrived to help carry the deer to the pickup.

Mitzi, Lady, Blue and I trailed a lot of other wounded deer over the years but after that incident I never let anyone besides myself take the final shot.

Another time, Arthur Von Rosenberg and his friend George Keller wounded a big buck on the two thousand-acre Von Rosenberg Ranch I was leasing at the time. They came by our house and wanted me to trail it up with my cowdogs. I took Mitzi and Lady to where the deer had been shot. There were a few drops of blood on the ground so Mitzi and Lady took off trailing the wounded animal. The deer wasn't wounded very badly and ran about a mile and a half into the Chaparosa Ranch before the dogs could stop him. We were running behind trying to keep up with the dogs and caught up to them right after they bayed the deer.

It was a big buck and Mitzi and Lady had it going round and round in a whitebrush thicket. I shot the deer in the neck and while we were gutting it the dogs went to a reservoir about two hundred yards away for a drink and a swim to cool off. There wasn't any brush between the tank and where we were working on the deer. In a few minutes we heard the dogs yelp and looked up to see Mitzi and Lady being chased by about fifteen mad javelinas. Their teeth were popping and their hair was all bristled up, making them look big and mean. I knew the dogs would come right to us so I told Arthur and George to step up in a mesquite tree.

The dogs were coming right toward us with the herd of javelinas in hot pursuit. Arthur and I were standing side by side and as the dogs and javelinas got closer we stepped up into two small mesquite trees and were about three or four feet above the ground. That is plenty high enough to be safe from a javelina. As the javelinas got almost to us I hollered at them. They stopped and started back to the reservoir, having chased the dogs away from their watering hole. As Arthur and I stepped down to the ground I kept hearing a noise behind me. I turned around to see George in the biggest mesquite tree in the thicket. He was about ten feet up and still climbing!

Arthur and I kidded George about climbing so high to get away from a bunch of javelinas. But George said it was his first experience with javelinas. He thought they were the meanest looking animals he had ever seen and they were coming right at him!

←―――――――→

One of my cowdogs had an encounter with javelinas that didn't turn out so lucky. One day my partner, Prince Wood, and I were putting out cottonseed meal and salt for our cattle. Our two cowdogs, Mitzi and Spot, were riding in the back of the pickup as usual. When we stopped at the first feed trough they jumped out and headed into the heavy brush to look for something more interesting.

While Prince and I were mixing the meal and salt we heard the dogs bark a few times and then some growling. All of a sudden we heard the sound of vicious fighting and the familiar popping sound javelinas make with their teeth when they're mad.

Prince and I ran to where the fighting was taking place to find Mitzi and Spot backed up in a big prickly pear cactus bush surrounded by about ten javelinas. Both dogs were trained not to chase javelinas so they probably just ran up on them and were attacked. The javelinas ran away when we came up. Mitzi was all right but Spot was cut under his throat and blood was running down his front leg in torrents. He was probably cut in the initial attack since he had bled so much and was already weak when we found him.

I laid Spot on the ground to examine the wound. The blood was gushing out in spurts so I knew his jugular vein had been cut by the razor-sharp tusk of a javelina. After a closer look I saw the jugular vein wasn't cut in two but had a slash about a quarter of an inch long parallel to the vein. By then Spot only had a few more minutes of blood left in him. I quickly pinched the slit closed between my thumb and forefinger to stop the bleeding. I told Prince to get a piece of the cotton string that was used to sew the sacks of cottonseed meal. I held the cut closed while Prince tied it with the string. I turned loose and the vein didn't bleed.

We finished putting out the feed and drove back to the house. I sprinkled some sulfanilamide powder on the wound to prevent infection and speed up the healing. I was afraid the cotton string would rot off before the slit in the vein healed completely. Evidently the string held long enough for the vein to heal because in about four weeks Spot was back working cattle.

Once I was driving by the Keller Farm west of Crystal City when I saw the biggest herd of javelinas I had ever seen. There were at least 25 in the herd and they were crossing the road coming out of the King Ware Ranch. A friend of mine named Juan lived on the Keller Farm and had some good javelina dogs. I had seen him plowing as I went by so I turned around and drove back to where he was plowing. I told him about the javelinas and he thanked me, saying he hadn't had a chance to go hunting in several weeks and was ready to go. As I left I saw him get off the tractor and head toward his pickup.

A few days later I saw Juan in town and he said, "Mr. Jack, I wish you hadn't told me about those javelinas. I took four of my dogs to hunt them and that big bunch of hogs killed three of my dogs and cut the other one so bad I've got him at Dr. Darter's (the local veterinarian)."

I felt bad about Juan's dogs. There aren't any animals in South Texas more dangerous than an angry javelina.

THE WORLD'S SLOWEST HORSE RACE

During the 1950s I was ranching near Batesville. One of my neighbors was Ed Cassin, whose family was among the first to ranch in the area. Ed was a top brush cowhand and one of the fastest men I have ever ridden with while roping wild cattle in the thick brush. His favorite brush horse was named Sundown and when Ed was riding Sundown very few cowboys could keep up with him in pursuit of a wild steer.

One hot day I was helping Ed round up cattle on his home ranch with about six or seven other cowboys. Ed told us, "I've got a big, long-legged 'Bramer' bull in this pasture that I sure want to get penned so I can take him to the auction. He's spoiled about penning and will try to jump out if we do get him in the corral."

I told Ed, "I wish I had known we were going to have to rope something in the brush because I would have ridden a faster horse. This one's a good cow horse but isn't very fast." My horse should have been fast because his dam was a stakes-winning Thoroughbred mare and his sire was a son of Mr. Underwood's good Quarter Horse; Dexter. But he was so slow I named him Possum.

That day Ed was riding a heavy-set gray horse that had a little Percheron blood in his pedigree. He was a good cow horse and could pull about anything from the saddlehorn. Because he was so stout Ed had named him Encino, which means "oak" in Spanish. Encino wasn't a fast horse and Ed regretted he was not riding Sundown because he would have a better chance of catching the bull if it took off running through the brush.

We rounded up the cattle in the pasture, including the wild Brahman bull, and started them toward the corral. Just as we got close to the corral the bull broke out of the herd like he was shot out of a cannon. Ed and I took off after him, tearing through the heavy brush as fast as our horses could go. The bull was fast and we ran at full throttle for half a mile before he started to tire and we began to close the gap. We were within twenty yards of the bull when he came to a barbed wire fence and jumped it without slowing down. There weren't any gates close by so Ed and I gave up the chase and headed back to the corral.

As we rode back together I began complaining about how slow my horse Possum was. Ed was upset that we didn't rope the bull and started griping about Encino being too slow to catch fast cattle. We began arguing about which of us had the slower horse, each insisting that our horse was slower than the other. By the time we got back to the corral we decided the only way to settle the issue was to match the horses in a race. The horse that lost was the slowest and therefore the winner.

Since we had to finish our cow work that day we agreed to race the next Saturday, running 200 yards on a dirt road in front of the Cassin Ranch headquarters. We bet a steak dinner on the outcome. By Saturday word of the horse race had spread through the little town of Batesville. A good many people from the area showed up to watch the "slowest horse" race. Quite a bit of money was wagered on which one of us had the slowest horse.

At race time we appointed someone to be the starter and two others to stand at each side of the road at the end to judge the finish. As we rode to the starting line I told Ed that since this was a race to prove which of us had the slower horse we would have to ride each other's horse. That way neither of us could be accused of "pulling" our horse to lose the race on purpose. Ed agreed, so we switched mounts and approached the starting line. The starter dropped his arm and the race was on.

We got off to a good start and I was really pouring the quirt to Encino, just as Ed was doing to Possum. The horses ran just about even for the first hundred yards and then Encino gradually pulled ahead of Possum and won the race by half a length. I won the race and a steak dinner by proving my horse was slower than Ed's horse. I don't know if Ed ever caught that bull.

Ed had a top leopard cowdog, named Dot, he used when he was roping out what we called remnants--the wild cattle that got away after a pasture had been gathered. I had several good leopard cowdogs myself but Ed's dog Dot was the best I ever saw for finding wild cattle.

One day Ed told me Dot had a litter of puppies a few days before and all but two had died. Ed really hated to lose those puppies because Dot's older offspring had become such good cowdogs. During that time South Texas was undergoing a severe drought and I was burning prickly pear cactus to feed about 500 mother cows on my ranch. A couple of days later I found four baby coyotes under a cactus bush. I immediately thought about Dot and the puppies that had died. I put two of the coyote pups in my pocket and at noon drove to Ed's ranch a few miles away.

I told Ed that a friend of mine, who ranched near Cotulla, had a good cowdog that had died after giving birth to puppies and only two had survived. I told Ed the mother had been bred to the best cowdog in that area so the pups should be tops. I told him that if Dot would raise them he could have one and I would keep one. Ed said he needed new blood in his dogs so he was glad to get them. The coyotes and the puppies were all about three days old so Dot readily accepted the new additions.

I didn't tell anyone around Batesville about the coyotes. About four weeks later I saw Ed at the Post Office in Batesville and he said, "Those puppies you brought me act different from Dot's pups. When I go into the barn, Dot's puppies waddle over to me wagging their tails. Those Cotulla pups just back away and growl at me."

I had to think fast to keep him from getting suspicious. I said, "Ed, they are just like their mother. She was one of those dogs that was not friendly with people and did not like to be petted. She stayed at the barn and around the horses. But when the cowboys saddled up and rode out in the pasture she was right there, ready to work cattle."

My story satisfied Ed and he repeated that he was looking forward to having some new cowdogs with different bloodlines. I didn't see Ed for several weeks after that conversation. Then one day, as I was driving into Batesville, I saw his pickup coming up behind me with its headlights blinking. I pulled over and stopped. Ed pulled up behind me, got out of his pickup and marched up to my pickup window. He said, "Jack, come get those &*#*@ coyotes!"

BUFORD THE JAVELINA

In the summer of 1970 our oldest son, Bob, caught a baby javelina as he was on his way to our ranch west of Crystal City. The little javelina was about two weeks old and Bob named him Buford. He soon became a favorite pet of the family. Buford and our cowdog, Lady, had a great time playing together. We never allowed pets in the house but Buford was an exception. Bob would often bring Buford into the house at night and he would sit on Bob's lap while they watched television.

In 1971 a lot of families moved to neighboring towns because of problems in the Crystal City schools. We put two mobile homes on some land I leased in Dimmit County near the Nueces River. We stayed there four nights a week so our kids could go to school in Carrizo Springs.

My wife, Evelyn, was also teaching in Carrizo Springs. Every Monday she would load the car with enough clothes for herself and the kids. On Monday afternoon I would go by our home to pick up my clothes and get Buford and Lady. I would open the tailgate of the pickup and they would jump in and we would head for the mobile homes. On Friday morning I would haul Buford and Lady back to our home in Crystal City.

Rosa Lopez had been house cleaning for us one day a week for several years. Buford accepted her as a family member and they had a pretty good relationship. That is, Buford would allow Rosa to come up to the house with out bristling up and popping his teeth as he did to strangers. We never had to worry about burglars or unwanted salesmen when Buford was around.

Our good friend, Billie Ruth Solansky, and her family lived about two hundred yards down the road from our mobile homes. In January of that first school year Billie Ruth called and said, "I sure hate to tell you this. Buford has been coming down to our house every night, running our dogs under the house, eating the dog food and causing the dogs to bark all night." Her son Johnny and our son Bruce had been friends all their lives and Billie Ruth knew how fond we all were of Buford. I apologized and told Billie Ruth I would take care of the situation.

The next day I took Buford up to our ranch about eight miles west of Crystal City. I put him out in a big pasture and drove off real fast. I looked in the mirror and saw Buford running after the pickup as fast as he could go. It really hurt me to go off and leave such a good pet but I knew it was the best thing to do. Everyone was sad about setting Buford free.

Gilbert Gonzalez was a ranch foreman on a ranch west of ours. He was a good family friend and knew Buford because he had once lived next door to us in Crystal City. One cold rainy night, about a week after I had taken Buford to the ranch, Gilbert called us at the mobile home and our daughter, Kay, answered the phone. Gilbert told Kay that Buford had been at the ranch corrals by the highway for three days. Gilbert said Buford would run after every car that went by and when it didn't stop he would go back to the corrals. Gilbert told Kay he was afraid someone would shoot Buford or he would get run over. I don't know why Buford went to those corrals since I had put him out about three quarters of a mile from the corrals and Bob had found him several miles from there.

Kay was a junior in high school and she loved Buford. She came into the living room in tears and said, "Daddy we have to go to the ranch and get Buford before he gets killed." So off we went to the ranch on that cold rainy night in January. I was secretly hoping Buford wouldn't be there when we arrived. It was still raining and pitch dark when we got to the ranch and I stopped in front of the corrals. I opened the door and started to call Buford but before I could finish saying his name, Buford, all wet and smelly, jumped into the cab of the pickup. He crawled under my legs and put his head in Kay's lap. We took Buford to our home in Crystal City and went back to the mobile home.

I guessed Buford missed his family, so after a few days he would go down the street and play with one of our neighbor's children. After about a week the neighbor called and said she was afraid Buford might get a little too rough and hurt one of the younger children. I took Buford back to the ranch and put him out on the lower end of the ranch almost two miles from the corrals. I called Gilbert Gonzalez and told him not to call if Buford came back to the corrals. We were sorry that we had to let Buford go but knew it was for the best.

I saw Gilbert about two weeks later and he said Buford had been at the corrals for about a week then disappeared. We never saw or heard of Buford again after that and hoped that he found a pack of javelinas to join. He was a great pet of the family for three years and we still come up with funny stories about Buford.

COWBOY WRECKS
Part II

One day two of my friends, Pete Simpson and Bubba Day, were going to help me move some cattle to another ranch. We needed two extra cow horses and Bubba suggested we get them at his ranch, a few miles south of Loma Vista. The three of us drove to there in my pickup and trailer to get the horses.

The Day Ranch had eight or nine horses in its remuda in a four hundred-acre horse trap. We drove out in the pasture where the horses were grazing, planning to catch the two we wanted without having to drive all of them to the corral. Since they were all gentle saddle horses we figured we could catch both of them with a bucket of feed.

I got out of the pickup and caught the first horse we wanted and tied the horn end of the rope around his neck. I used a horse knot to tie the rope, one that wouldn't slip and choke the horse. I was going to put the loop end of the same rope around the other horse's neck, but Bubba came up and caught it instead. Since these ranch horses hadn't been loaded in a trailer very often I decided to lead my horse into the trailer. I had the coiled rope in my hand as I led him into the trailer. When the horse jumped up into the trailer the sudden movement of the trailer scared the horse and he started backing out of the trailer in a hurry.

I stepped out of the trailer with the horse and accidentally dropped the loop of the rope in front of me. The horse kept backing up real fast and all of a sudden my feet were jerked out from under me. I had put my left foot through the loop and it tightened up just below my knee. When I hit the ground it scared the already spooked horse. He wheeled and started running away, dragging me with him. All the other horses started running, which caused my horse to try to keep up with them.

This was during the drought so the dust was really flying as I was being dragged along. The pasture had lots of brush and it was a bumpy ride, but I managed to reach in my pocket and get my knife out. The thought going through my mind was that my knife wasn't very sharp and I wish I had sharpened it more often. I managed to get the blade open and just as I was reaching for the rope the horse dragged me through a big bush. The knife slipped and I thought I had dropped it. When I looked at my hand I could see about an inch of blade sticking out of my fist. I caught the rope at my knee and made a cut with the end of the blade. Nothing happened, so I cut a little harder and one of the three main strands started unraveling.

All this time Pete and Bubba were trying to figure out what to do. Bubba remembered the 30-30 Winchester rifle in my truck and ran to get it. He thought the only way to save my life was to shoot the horse, which would have been a pretty long shot.

By then the horse was really moving fast and I was hitting lots of brush and prickly pear and stirring up a lot of dust. I made another cut and the second strand unraveled. I felt a little better because I thought that if I hit a big solid bush the third strand might break. I reached for the third strand and cut it. The horse and I parted company. I had been dragged about three hundred yards and was really skinned up. My Levis and shirt were torn and I was full of assorted thorns. My leg really hurt, but I was very luck to be alive.

The horses ran on into the corrals and stopped. Pete and Bubba caught the two horses we wanted and we loaded them in the trailer and headed back to my ranch. My wife, Evelyn, picked the prickly pear thorns out of me and I took a bath. She drove me to Uvalde to have my leg X-rayed, which showed a small chip out of the leg bone just below the knee. Pete and Bubba went ahead and worked my cattle that day. Neither one of them had a knife so they detoured by Batesville and each bought a new pocketknife. I keep my knife sharpened and never go anywhere without it.

Another memorable wreck occurred in the 1960s on my ranch in Zavala County, Texas. I was riding my good gray stud, Bandy King, in a brushy pasture roping out some big wild steers. The steer I was after had wide horns so I built a big loop. We were running as fast as a horse and wild steer can run, angling toward a barbed wire fence with big cedar posts. I knew the steer would turn at the fence and that's when I planned to rope him.

The wind was blowing a gale behind me as I was swinging my big loop, ready to throw when the steer made his turn. I was making my last swing when the loop caught the top of a cedar fence post. I was using a nylon rope tied to the saddlehorn and we were flying. I immediately let go of the rope and tried to pull up Bandy King. Before he could stop the slack ran out and the rope tightened with a sudden jerk. Bandy reared up high on his hind legs and then suddenly shot forward at a forty-five degree angle to the ground.

His momentum had pulled the big cedar post completely out of the ground and free of the barbed wire. Just as Bandy landed on all fours the post flew by us like it was shot out of a cannon, with the rope still attached. The fence was on our right and the post passed us on the left, barely missing me. The rope wrapped around Bandy's hind legs and when the post reached the end of the rope it jerked him halfway around. I don't see how Bandy kept from falling but he managed to stay on his feet. If the post had hit me it could have easily killed me. The next morning both of Bandy King's hocks were swollen as big as my head. It was about two months before I could ride him again.

←——————→

One time I was helping my neighbor catch two bulls that his cowboys hadn't been able to pen. I took my best horse, Bobo, and another young horse I had been riding for about six months. The young horse was about three years old and I had put lots of miles on him, but hadn't started him on bits because he handled so good with a hackamore. I had roped several cattle on him so he knew how to handle himself when you needed to rope something.

That morning we a caught one of the bulls, a big crossbred, while I was riding Bobo. We had a pretty long run catching the bull and after we hauled him to the pens I noticed Bobo was a little gimpy. That afternoon I rode my young horse to hunt for the other bull, a Brahman with big horns and a big hump on his back. The man who owned the bull rode down one draw and I went over a low hill to ride down another draw. We were pretty sure the bull was hiding in one of the two draws since they were both real brushy.

The pasture we were in was about three thousand acres, big even for South Texas. I had ridden about two hours when the bull suddenly jumped up in front of me. I took off after him and we ran a good ways before I could get close enough to rope him. I built a big loop because the bull's horns were so big. When I threw it the loop not only cleared the horns but also the Brahman's hump. The rope was around the bull's shoulders rather than its neck and tied tight to my saddlehorn. Any cowboy who has roped a lot of Brahman cattle will tell you when you catch a big Brahman bull behind his hump the bull just about always wins. He dragged us a long way before my horse could stop him. There I was, sitting on an inexperienced young horse with a hackamore instead of a bridle, tied to very mad 1600-lb. Brahman bull. Believe me, there are a lot of better places to be.

The bull suddenly wheeled and charged. He hooked at us as he ran past and I had to turn the horse quickly, so he would be facing the bull when he hit the end of the rope. We survived the first big jerk and were ready when he charged at us again. The big old bull was really mad by then. He started chasing and trying to hook us. The biggest problem in running from a bull is keeping the rope from getting under your horse's feet, which would really be a wreck. I tried to get close enough to a mesquite tree to wrap the rope around and hold the bull but twice the trees broke off at the ground. Things were not getting any better.

All this time I kept hollering for the owner of the bull to come help me, but the wind was blowing so hard he couldn't hear me. We played cat and mouse with the bull for two exciting hours. Sometimes the bull would stand and look at us for a good five minutes before he charged. Other times he would just wheel and drag us a good ways before we could stop him. I noticed he always shook his head five times just before he charged. That got to be a pattern; shake his head five times and he was going to charge; shake his head two or three times and he would turn and run.

The bull's owner finally heard me and came to help. He roped the bull and we tied him to a big mesquite tree so we could haul him to the corrals. Those two hours gave my young horse a year's worth of experience.

One day my partner, Prince Wood, and I were helping our neighbor, Dave Leibman, rope out several old Brahman bulls that were hard to pen. I roped one of the bulls around the neck and one of Dave's ranch hands, Jose, rode up and threw a rope around the bull's horns as he charged me. The bull hit the end of Jose's rope before Jose could turn his horse to face the bull, jerking his saddle sideways and throwing Jose to the ground. Jose's horse started pitching which pulled the saddle, still tied to the bull, back toward his flank. The horse then started running and pitching in a circle.

There I was, holding the bull with a rope tied to my saddlehorn, while Jose's horse was coming right at me. Jose's rope was longer than mine, so his horse came around behind my horse. When his rope tightened around my horse's hind legs both horses hit the ground. I was thrown down between them and Jose fell off to the side. That's when the bull decided to join the pile and charged. Jose's horse was trying to get up when the bull arrived at full speed, knocking his horse back down. I managed to crawl free of the pile and ran toward Jose, who was putting some distance between him and the wreck.

About that time Prince came up and roped the bull. He held him away from the horses so they could scramble to their feet. Jose and I caught our horses. I got back on my horse to help Prince hold the bull while Jose took his saddle off his horse. Nobody said much as we loaded the bull in a trailer to haul him to the corrals. We were lucky that everyone, including the horses, survived.

BICYCLE WRECKS

Our four kids would often ask, "Dad, what did you do for fun back in the olden days when you were kids? You didn't have television. Life must have really been boring."

Life wasn't near as boring as they thought. Here is an example: In the early 1930s when I was eleven we were living on our ranch south of Santa Anna in Coleman County, Texas. We kept begging our parents for a bicycle. They couldn't understand why we wanted a bicycle since we had plenty of horses to ride anytime we wanted to saddle one up. Most kids our age would have quickly traded their bicycles for a horse to ride.

To our surprise we got a pretty red bicycle that Christmas. We were thrilled and could hardly wait until we could go outside and ride it. With only one bicycle and three eager boys there were some heated arguments as to whose turn it was to ride and for how long. One Sunday we got home from church, changed clothes and went outside to ride the bicycle while Mother fixed dinner. We agreed we would take turns riding the bicycle once around the house then let the next one ride.

Hank was the oldest so he insisted on going first. He took off while Tom and I waited our turn, but when he came around the house he kept going. He came by us a second time and we started hollering for him to stop and let us ride. I saw he wasn't going to stop after the third round so I picked up a broken broom handle and threw it like a spear toward the front wheel.

The broomstick went between the spokes of the front wheel and the fork that held the wheel. Hank was really pedaling fast by then and the broomstick broke nearly all the spokes, causing the bike to flip, throwing Hank to the ground. He wasn't hurt, but was mad about the broken bike. Tom and I explained to Mother that Hank wasn't being fair in sharing the bicycle. We didn't know what Dad would say about the damaged bicycle. He wasn't too happy but after a few days took the wheel to town and had the broken spokes replaced.

After a while the new wore off and there weren't many arguments about who rode the bike. Everyday one of us had to ride about a mile down the county road to get the mail. One windy March day I was on my way home, riding a newly broken young horse, when I saw Hank getting the mail out of the mailbox. He was riding the bicycle so we rode along together for a ways. We were headed into the wind and Hank was having a hard time pedaling. About three-quarters of a mile from the house he asked, "Why don't you pull me with your horse?" I was riding the horse with a hackamore and always carried a lariat rope tied to the saddlehorn. I tossed the loop to Hank and started trotting the horse, pulling the bicycle with Hank toward the house.

We were going along just fine when Hank hollered "Go faster!" The young horse hadn't seemed spooked when were travelling at a regular trot but when I kicked him into a lope he noticed the bicycle behind suddenly speed up. He began to cock his right ear back and I could tell the horse was really spooked. I started trying to slow him down with the hackamore reins but the harder I pulled the faster he went.

We were really beginning to move and I thought Hank would turn loose of the lariat so I could get the horse under control. When I looked back to see why he was still attached I was shocked to see that Hank had put the loop over the handlebars and it was pulled tight around the steering column. There was no way he was going to get that rope off the handlebars. All I could do was sit up there on the horse and pull hard and holler "Whoa!" Hank was hollering, "Stop! Stop!" and putting on the brakes. That didn't help a bit and by then the horse was running as fast as it could go with the bicycle in close pursuit. I knew the horse wouldn't stop until we got to the house.

There were two obstacles we had to survive before we got to the house. The first was a sharp ninety-degree turn in the road about two hundred yards from our house. I figured the horse could make the turn but there was no way the bicycle could. The second obstacle was a gate we had to turn into just before the house. It had two solid mesquite posts on each side and as we were going parallel to the fence there was no way Hank would be able to make the turn and miss both gateposts. Of course he would have to survive the curve before worrying about the gate.

As we approached the sharp curve, I glanced back and saw Hank bent over and hanging on for dear life. As we went around the curve the bicycle skidded sideways and went down, sending Hank rolling at least twenty or thirty feet. The empty bicycle was bouncing from one side of the lane to the other. I pulled the horse close to the far side of the lane just before he turned to go through the gate. The turn wasn't wide enough and the bicycle hit the first gatepost hard and whipped over to the other post, completely wrecking it. The horse ran to the corner of a big corral and stopped. What was left of the bicycle kept coming and hit him in the rear. The horse finished off the bicycle by kicking it all the way to the end of the rope.

    I took the rope off the saddlehorn and rode back down the road to check on Hank. He was skinned up some but otherwise OK. Hank didn't have much room to chew me out since he realized he probably shouldn't have put the rope over the handlebars. Later Dad looked at the wrecked bicycle and at us, but didn't say anything. Needless to say we didn't ask him for another bicycle.

THE CLOTHESLINE WRECK

One day in 1963 I was hauling a yearling steer to the Carrizo Springs locker plant. I had been feeding this steer for about two months and was having him processed for our freezer at home. I had my two-year-old son, Bruce, with me that day. The locker plant was right in town.

After I backed the trailer up to the unloading chute I saw it wasn't quite close enough. The trailer needed to be a few inches closer to the chute so the tailgate would hit the side of the chute and keep the steer from pushing it out and escaping. I started to get in my pickup and back up those few inches but then decided not to. The steer was fairly gentle and I thought I could hold the tailgate and the steer would go on into the chute. I opened the tailgate and held it as the steer stepped out of the trailer. Just as he started to go into the chute he hooked at me through the tailgate; pushed it open and took off to explore Carrizo Springs.

There I was, with a pickup and trailer, a two-year-old boy and a loose steer running around Carrizo Springs. A good friend of mine, Gene Allen, lived just east of town where he kept his good roping horse, Chupadero. I called Gene from the locker plant office to see if he could help me catch the steer. His wife, Vida Jo, answered and told me Gene was hauling a load of cattle, but to come get Chupadero. Vida Jo said she would follow the steer in her car while I was saddling the horse. She arrived in a few minutes and took Bruce, who started crying, in her car.

I rushed out to the Allen's house, saddled Chupadero with Gene's saddle, got his rope and loaded him in my trailer. It took me a little while to find Vida Jo, who was following the steer as it traveled east along a brushy creek that runs through the south part of Carrizo Springs. I unloaded Chupadero and started after the steer. I managed to turn him north close to where another friend, Cecil Hancock, lived. I made a loop and got ready to rope the steer as he came out of the brushy creek. He came by me along a chain link fence in someone's back yard. I threw a long loop, which the steer jumped right through, catching his left hind leg.

There was a clothesline in the back yard where I caught the steer. It was made of two cedar posts with cross boards at the top and three wires to hang clothes. Both cedar posts had a brace wire running from the top of the post to the ground to keep the post in place. The lady of the house had apparently just washed her clothes and hung them on the clothesline.

As the rope tightened on the steer's hind leg he ran past one of the brace wires. The rope was over the brace wire and every time the steer kicked, the clothesline would jerk, sending freshly washed clothes flying. While this was going on the lady was standing in her door looking on. I tried to back up Chupadero to get the rope off the brace wire but we were in a corner. I rode forward, keeping the rope tight so the steer wouldn't step out of the loop and take off again. We were within a few feet of the back door and I said to the lady, "I'm sorry. I'll be back." She didn't say anything, just looked at her clothes lying in the yard.

Cecil Hancock happened to be out in his yard when all this was going on. His stock trailer was hooked to his pickup in his back yard. The steer was heading that way so I just followed, keeping the rope tight enough so it wouldn't come off his leg. By then the steer was pretty mad and looking for someone to hook. He saw Cecil and headed right toward him. Cecil opened the tailgate and got behind it as the steer approached. I kept letting the steer go forward until he moved close to the back of the trailer. Cecil swung the tailgate around and forced the steer into the trailer and latched the gate. I took the rope off the saddlehorn and pitched it in the trailer.

I rode across the street to the lady's house, got off Chupadero and started picking up clothes. The lady came out and said, "Don't worry about them. I'll pick them up." I told her I was sure sorry for what had happened and I would take the clothes to the laundry to have them washed. I started to give her some money for the damage I had done. She wouldn't take the money and told me not to worry about it. "You have enough problems as it is."

I rode back across the street and Cecil said he would take the steer to the locker plant. I rode Chupadero back to my pickup and loaded him in the trailer. As we unloaded the steer at the locker plant I made sure Cecil's trailer was close enough to the unloading chute this time. I thanked Cecil, took Chupadero back to the Allen's house, got Bruce, who was all smiles by then, and drove home to Crystal City.

OIL FIELD TALES

Natural gas was discovered on the Kingsbery Ranch, south of Santa Anna, Texas, in 1914 and was piped into Santa Anna in 1916. This was the first natural gas to be used in homes in that part of Texas and most homeowners were afraid of it. It took a lot of public relations work by the owners of the gas system to convince the people of Santa Anna that gas was safe for home use.

To commemorate the coming of gas to Santa Anna a gas street light with a large glass cage was installed in the center of the main street running through town. My aunt, Merle Kingsbery, was in the first graduating class of Southern Methodist University that year. Since her father, H.W. Kingsbery, owned the ranch that produced the gas and was also president of the local bank, Merle was given the honor of lighting the street light.

A large crowd of people assembled around the street light, listening to speeches about the wonderful benefits of having natural gas in their homes and how safe it was. Finally it was time to light the gas. Aunt Merle stepped up on a stool and a lighted torch was handed to her. A signal was given to a person across the street to turn on the gas. This person got the signal a little too soon because when Aunt Merle stuck the torch in the window of the glass cage it blew up with a big flash. Luckily no one was hurt but it certainly wasn't a good way to convince people that natural gas was safe. Incidentally the people of Santa Anna are still using gas from the Kingsbery Ranch.

In 1929 an oil well was completed on the Kingsbery Ranch near the site of the 1914 gas well that was still producing gas. The well was 1100 to 1200 feet deep and produced about 30 barrels of oil a day. A number of wells were drilled in the same area after that and they all produced 10 to 30 barrels a day.

My younger brother Tom and I were about ten and eleven years old and we liked to snoop around the drill site after a drilling rig had been moved to another site. The main thing we were looking for was lead or babbit (a soft metal alloy) that was used around those drilling rigs. Tom and I would melt the lead and pour it into molds to make all types of different things.

The oil field was about two miles from Uncle Carroll Kingsbery's house. One day Tom and I had ridden two young horses we were breaking to an oil well site that had just been completed. It was Sunday and the rig hands had the day off. We tied our two broncs to a big tree and started walking around the well site. There was a pipe coming out of the well that had a two-inch pipe screwed into it at a right angle about four feet above the ground. The two-inch pipe had a valve on it and was open at the end pointing toward the slush pit.

Since the drillers had just finished the well they hadn't told Dad whether it was a gas well, an oil well or a dry hole. My curiosity got the best of me so I started opening the valve. There wasn't anyone within two miles of the well except Tom and I so I didn't think I would get in trouble for "checking" the well. At the first turn of the valve natural gas started spewing out the end of the pipe. It made a great sound so I opened the valve some more. The more I opened it the louder it got, creating a deafening roar. Suddenly Tom started hitting me on the back and screaming for me to close the valve. I closed it and looked around in time to see our two horses running at top speed toward Uncle Carroll's house. The sound of the gas had terrified them. They broke their hackamore reins and took off to the barn. Tom and I had to walk two miles to get our horses. The well turned out to be a good gas producer.

By the early 1930s the oil company had completed twelve wells. There wasn't any electricity in the area to pump the oil out of the wells. The oil company constructed a large tin building as a pump house near the center of the wells. They installed a huge one-cylinder gas engine with two large flywheels in one end of the pump house. At the other end of the building they installed a huge horizontal wheel on an axle that came out of the concrete floor. This wheel was about twenty feet in diameter, almost twenty inches thick and was about four feet above the floor. A smaller wheel was mounted on one of the big wheel's spokes a few feet from the center.

The big wheel was turned by a heavy-duty drive belt, eighteen inches wide and seventy-five feet long, running from the drive pulley on the big gas engine. The small wheel had about fifteen holes around the edge attached to sucker rods. The sucker rods went out of the pump house in all directions to the pump jacks on the various wells. The engine turned the big wheel, making the small wheel go around in a circle. This made the sucker rods slide back and forth, pumping all the wells at the same time.

The oil company had a man, called a pumper, who lived on the lease and took care of the machinery. Some of the wells were nearly half a mile from the pump house and the pumper would put 2X8 boards on the ground for the sucker rods to slide. The rods would move back and forth about three feet. Some of the rods from the pump house would travel through slits cut in 2X4 boards and fitted in two-inch pipes driven in the ground. Some of the rods would be about a foot above the ground.

When we were working or checking cattle in the oil field pasture we had to ride through the maze of sucker rods and oil wells. Most of the time the grass would be about six inches tall. The horses wouldn't see the moving rods under them until they were right on top of them. When they saw the snake-like movement of the rod they would throw a fit. It was hard to get them to cross the almost hidden moving rod. It was even harder to get a horse to step over the rods that were a foot above the ground.

One Saturday afternoon when Tom and I were ten and eleven years old we were returning from checking the cattle when we rode by the pump house. The engine was running and the rods were moving back and forth. The machinery fascinated both of us and we had gone in the pump house many times to watch it. As we stood by the big wheel we noticed there were hundreds of jumbo grasshoppers in the pump house. They were about three inches long and a lot larger than regular grasshoppers and had no wings. I picked one up to look at and for some reason pitched it toward the big wheel just where the belt came in contact with the wheel. The grasshopper made a loud popping noise as it was squashed between the belt and the wheel. Tom picked up another grasshopper and pitched it on the belt. It made the same loud pop so we both started throwing grasshoppers onto the belt.

In a little while I noticed the belt slowing down. I looked back at the engine and saw the belt beginning to smoke where it went around the drive pulley on the engine. I realized the mashed grasshoppers had made the belt slick and that it was slipping. I ran to the engine knowing it had to be stopped, but I didn't know how. When I got to the engine the belt had completely stopped and was really beginning to smoke. The only thing I could think of to stop the huge engine was to jerk the wire off the spark plug. That stopped the engine.

We were two scared young boys. The pumper had gone to town, so we got on our horses and rode home. We couldn't get up enough nerve to tell Daddy so we never told anyone. A few days later we rode over to the oil well pasture to check the cattle and the pumping unit was running normally. As we rode by, the pumper waved at us. We talked to him lots of times after that and he never mentioned anything about the engine stopping. Tom and I hoped he hadn't connected us to the grasshoppers.

About fifteen years later, Tom and I were drinking coffee in the Donham Cafe in Santa Anna when the pumper came in. He seemed glad to see us and sat down with us. Tom and I had both been in the Army and then away at college and hadn't seen him in a good many years. He had retired from the oil field and we talked about things that had happened while we were away. As he was ready to leave he turned and said, "Boys, I wonder if there are a lot of those big old grasshoppers in the pump house on the oil lease?" He kind of smiled, got up and walked out of the café. He knew all the time but never told our father.

In 1937 Tom and I were riding our good calf-roping horses in the oil well pasture. We rode up to one of the older well sites that was still in production. The oil well crew would have to pull the sucker rods out of the well when the pump leathers needed replacing. They had a mast made from six-inch pipes that was forty feet tall with a pulley at the top. The mast would be dragged to a well site when the sucker rods needed pulling. The crew used a big Fordson tractor with a large winch on the front to pull the mast to the well site and raise it. Four guy cables held the mast up.

The crew had pulled the sucker rods the day before at this well and the mast was still standing. We rode up to the well, got off our horses and walked around the area. The four guy cables were tied to earth anchors that were deep in the ground. I noticed one of the cables was loose and I thought it should be tighter so I started turning the turnbuckle by hand. The cable was loose enough for me to unhook it from the earth anchor so I could take up the slack faster. After I had taken up part of the slack I went to hook it back in the eye of the anchor and finish tightening it. When I tried to hook the turnbuckle into the anchor it was too short. The heavy cable was about fifty feet long and I was really straining to hook it back in the anchor. I pulled hard again and realized the hook at the end was farther from the anchor than the first time. I was suddenly in shock as I could feel the guy cable pulling me along the ground as I was pulling as hard as I could. The mast was beginning to fall!

I hollered for Tom to get out of the way and almost fainted when I realized my horse, Queenie, was in the path of the 1000-pound falling mast. She was my top roping horse and had her head down grazing. Luckily the mast landed a few feet behind her. What was intended to be a good deed almost turned into a disaster!

One morning I was riding a young horse by a drilling rig that was about three miles from the other wells. It was called a "wildcat" since it was in an area that hadn't been drilled. There were three or four men on the drilling platform deck running the operation. Back in those days the drilling was done with cable tools. Water was run in the hole and a heavy steel bit held by a cable was lowered into the hole. The drilling rig ran with an up and down motion that loosened the soil and made thin mud when mixed with the water. This churning procedure would go on for twenty or thirty minutes and then the bit would be pulled out of the well by the cable. A bailer would be lowered into the hole to bring the mud out and put it into a slush pit. The bailer was made of heavy steel pipe, about twenty feet long, with an attachment at the bottom of the pipe to hold the mud in the bailer as it was lifted out of the well.

The men had been drilling about a week and were down around seven hundred feet the day I rode up to watch. As I sat on my horse the men pulled the bit out of the well and hooked the cable to the bailer and lowered it into the well. I watched them bring it out full of mud and empty it into the slush pit. They put the bailer back into the hole for another load of mud. The bailer had just enough time to reach the bottom of the well when I heard a rumbling noise coming from the hole. The drillers started running off the drilling platform and jumping to the ground. The bailer had hit a pocket of gas and immediately there was a great roar as the gas come out of the hole.

My horse wheeled and started running away from the rig. I was trying to hold him and at the same time look back to see what was happening at the well. The roar got louder and in a moment the bailer came flying out of the hole. I wasn't having much luck holding my horse, but looking back over my shoulder I saw the big bailer flying through the air and bounce off the top of the rig with the cable still tied to it. That was certainly a sight to see as the bailer and cable went several hundred feet into the air. I finally got my horse stopped and by then the roaring in the well had about died down.

The men were walking back to the rig as I rode up. They said it was lucky the gas didn't catch on fire. The blowout did mess up the well so the oil company didn't try to go any deeper. About twenty years later another oil company moved a rotary rig near the site and drilled down three thousand feet but didn't hit any gas or oil.

SQUEEZE CHUTES, SADDLE RACKS & GUN SAFES

Crystal City, Texas is famous for being the spinach capital of the world, symbolized by a statue of Popeye in front of City Hall. It is less well known as the birthplace of the cattle squeeze chute, a device that holds cattle immobile so they can be branded, de-horned and doctored. Until the squeeze chute was invented, working cattle was a battle between man and animal. Often the animal won. Squeeze chutes made ranching a lot less dangerous.

Mr. Clay Shearer developed the squeeze chute during the mid-1930s and patented it in 1939. I still have the scale model that was taken to Washington D.C. to get the patent. Mr Shearer and Mr. A.C. Mogford formed a partnership to manufacture and sell squeeze chutes. Mr. Mogford later bought out Mr. Shearer and changed the name to Mogford Industries. Over the years the company sold hundreds of the famous Crystal City Squeeze Chutes to ranchers and veterinarians all over the Southwest. There were several of the chutes on the ranches I leased.

One day in 1963 I hauled one of the chutes into Crystal City to have some work done on it. Mr. Mogford and I were good friends. He had graduated from Texas A&M (Class of 1922) and was also a schoolteacher like my wife, Evelyn. While his men were unloading the chute from my trailer Mr. Mogford walked up and we talked about the repairs. After a few minutes he said, "Jack, why don't you buy this company? My wife and I don't have any children to leave it to and since you're an engineer you could improve the chute."

I had majored in mechanical engineering at A&M but decided I wanted to be a rancher and ended up with a degree in animal husbandry. Mr. Mogford added, "You know every rancher in South Texas and could sell a lot of chutes." My answer to Mr. Mogford was "What in the world would I do with a squeeze chute company?"

About a week later I returned to pick up my chute. As the men were loading it, Mr. Mogford asked me if I had decided to buy the company. I said, "Gosh Mr. Mogford, I haven't even thought about it since you mentioned it last week." I looked around in the building at the various machines and all the different materials. Then I asked him what he would take for the company. He thought for a few minutes and told me what he wanted for it. He said he would either sell me the building or lease it to me. I thought for a few minutes, shook hands with him and told him I would take the company. I took the chute back to the ranch and worked cattle in it that afternoon.

At the time I was running about 400 mother cows and putting several hundred yearlings on irrigated oats during the winter. I had two Quarter horse studs and was breeding other peoples' mares to them as well as my own ranch mares. I was chairman of the of the Zavala County Agricultural Stabilization and Conservation Service committee, vice-chairman of the Soil Conservation Service committee and president of the Crystal City Lions Club. Evelyn taught school and we had four kids, ranging in age from eleven to two. They helped with the ranch work when they could but I did most of the work and had a really busy schedule.

That evening we were all eating dinner when I looked at Evelyn and said, "Honey, guess what I did today." Evelyn answered, "What?"

"I bought Mogford Industries." After about a minute, Evelyn said, "Jack, you really need Mogford Industries. Between twelve o'clock at night and four in the morning you don't have a thing to do."

A few weeks later Mr. Mogford and I closed the deal and I was the owner of the world's first squeeze chute company. I changed the name to Kingsbery Manufacturing Corporation. Running a company was a different world from cattle ranching and a real challenge. I used my experience as a rancher and my education as an engineer to make several major changes on the squeeze chute. I replaced the side squeeze lever with a rotating drum and rope to eliminate the danger of the big pipe lever flying up and hurting someone. I never liked to walk around the lever to get to the head of the animal. The rope control also allowed the chute to fit in a narrower space.

The original Crystal City Squeeze Chute was made mostly out of wood. Later models were made mostly of steel with wooden side panels that dropped down for access to the animal. When I was working cattle it always seemed that a pipe or a board was in the way. I changed the sides to vertical pipes that were hinged at the bottom. You could drop down any or all of the pipes and get to any part of the animal. Veterinarians loved this new design and it was really handy for getting a newborn calf to suckle its mother.

About the time I bought Mogford Industries, Texas had introduced a statewide program to make Texas a brucellosis-free state. Brucellosis is an infectious disease that causes cows to abort their calves and causes severe financial losses for ranchers. The Texas Animal Health Commission was in charge of the program and they had state veterinarians "bleed" cattle (take blood samples) and send them to laboratories. Very few small ranches and farms had facilities to hold grown cattle, so the state veterinarians were issued a stanchion-type wooden headgate made of oak. These headgates weighed over 300 pounds and it was a real chore to unload them from a car or pickup and wrestle them into place.

Two of the veterinarians came to see me about making a lightweight headgate. I told them I could make one out of structural aluminum that would weigh under 100 pounds and cost about the same as the heavy oak headgates. They asked me to make two of the headgates to test. If they worked well they would get the state to buy a bunch of them. They were going to take the first one to a ranch above San Antonio where they were going to bleed about 30 big Hereford bulls.

It took two weeks to make the first headgate and it ended up weighing 94 pounds. The two veterinarians picked it up and carried it to the ranch to bleed the Hereford bulls. I knew it would be strong enough and work well because I had tested it on three of my older Beefmaster bulls that weighed about 2,000 pounds each. That evening one of the veterinarians called me and said the headgate worked perfectly and to go ahead and build the second prototype. I built it the following week and sent it to a state veterinarian in North Texas.

The two headgates were used by a number of state veterinarians for a few weeks at a time, then passed on to other veterinarians. They sure hated to give up the Kingsbery headgate and go back to the heavy wooden ones. After the veterinarians had used the aluminum headgates for several months they asked the Animal Health Commission to replace the heavy oak headgates. The Commission said they were short of money and kept putting off sending me a purchase order. The two veterinarians who asked me to build the headgates were very apologetic and kept telling me that the state would come up with the money to pay for the first two, but I never did get paid.

I did however get calls from ranchers all over Texas who had seen the first two headgates. They wanted a Kingsbery Aluminum Headgate but we were behind on production of our other products so I never pursued that opportunity. After about two years one of the headgates was returned to me and I still have it. I never did get the other one back.

One of the hardest ranch jobs is working calves. It is hard work to hold a calf while it is being vaccinated, castrated and branded. I developed a tilting calf table to hold a 200-lb to 500-lb calf on its side. Two people could work calves all day with my calf table. We sold over 500 Kingsbery Calf Tables to ranches all over Texas and beyond. The King Ranch ordered one, along with one of my squeeze chutes, for their ranch in Spain. We put the calf table inside the squeeze chute so neither one of them had to be crated for overseas shipment.

Later I developed a revolving saddle rack called a Saddle Susan. It had three revolving racks, one held six saddles, above it was a saddle blanket rack and above that was a rack with hooks for bridles and other tack. The Saddle Susan took up a lot less space than usually required to store six saddles. Using the same principle as the Saddle Susan I designed a revolving display rack, called a Display Susan, for displaying all types of merchandise. I patented both inventions and sold them to ranchers, horse farms and retail stores all over the country. Sears Roebuck & Company bought three hundred Saddle Susans to display saddles and other riding equipment in their stores.

One interesting aspect of the revolving display rack I patented was the letters I got from lawyers all over the U.S. telling me that a certain company was infringing on my patent. My patent covered almost every type of revolving display racks, such as those you see for watches and paperback books. The lawyers said that if I gave them the OK to sue companies like Timex and Signet Books I would get half the amount recovered. I don't like litigation so I never answered their letters.

Kingsbery Manufacturing also made galvanized road culverts, called tinhorns, for many years. We had a big powerful machine that would roll heavy-gauge flat galvanized sheets into whatever diameter that was needed for the culvert. The rolled sections would be riveted together with a heavy-duty riveting machine to make the proper length. We made culverts from 12 inches in diameter to six feet in diameter and sold them to county road departments all over South Texas.

I also developed rectangle and triangle stall feeders for horses that could be hung on the side or in the corner of a stall. These heavy-duty steel feeders had a hayrack in the top part with a trough for grain in the bottom. They were ideal for feeding alfalfa hay. The leaves would fall through the hayrack into the grain trough and not be wasted. We sold thousands of these feeders all over the United States, shipping many of them by train because Missouri Pacific Railroad had a depot down the street from the company.

One time a rancher in Colorado and a rancher in Illinois each ordered two of the rectangle horse feeders. We shipped them by Missouri Pacific the same day. Each feeder was properly tagged, with two going to Colorado and two going to Illinois. About five days later I got a call from the Colorado rancher who told me the two feeders had arrived in Colorado but one had a tag with his name on it and the other had a tag with the Illinois rancher on it. Missouri Pacific wouldn't give him the Illinois feeder. I called the Illinois rancher and he said the same thing happened to him. I then talked to the train station manager in Illinois, explaining that the feeders were exactly alike and asked him to give the Illinois rancher the Colorado-tagged feeder. I told him I would call the station manager in Colorado to do the same thing.

I ended up making about eight phone calls trying to trade feeders but could never get Missouri Pacific to cooperate. In the end Missouri Pacific shipped the two feeders in opposite directions across the country. It took the ranchers about a month to get their feeders. No wonder railroads fell on hard times.

With Kingsbery Manufacturing I got to put my ranching experience into improving and inventing cattle handling equipment. In 1971, when someone broke into our home in Crystal City and stole our guns and other items, I used that experience to create a whole new business. The thieves took eight rifles and shotguns and two pistols. Most of them were valuable and several of the guns had sentimental value. It really hurt to lose those guns, some of which had been in my family for years, and I didn't want it to happen again. I looked in all the gun magazines for a safe-type gun cabinet. One company made a little chest-type safe out of lightweight steel that didn't look very secure.

I decided to design my own gun safe and drew out a blueprint for one. I took the inspection plate off the inside of a big warehouse safe to see how the locking mechanism worked. I decided to go to San Antonio to buy a combination safe lock and get some input about building a safe. I took a load of calves to Union Stockyards in San Antonio and later, wearing my work clothes, went to the biggest safe company in San Antonio. I asked to see the owner and told him what I planned to do, asking for his advice. He indicated he didn't have time to talk, really snubbing me. An older man was sitting at a desk nearby and heard the conversation. As I left he got up and followed me to the lobby. He stopped me and asked me to go with him to another part of the building. He showed me several combination locks and gave me some good information about which lock to use. I bought a lock from him and thanked him for the advice.

The following week my general manger, Phil Alcozer, and I took my blueprint and started building a safe. Phil was so knowledgeable about types of metal, welding and metal fabrication that his input was a major factor in building the first Kingsbery Gun Safe. The first safe worked well and looked good. I started advertising it and showing it at gun shows around the state and before long our gun safes were outselling squeeze chutes. We offered several different sizes and would custom-make any size gun safe. Harking back to the Saddle Susan we even designed a gun safe with a revolving gun rack.

We shipped Kingsbery Gun Safes all over the U.S. and to several foreign countries. We air freighted three safes to the island of Borneo and two safes to Banjul, Gambia in West Africa. One customer wanted to know if we could put two combination locks on a safe so that it would take two people to open it. I told him we had never built one like that, but I was sure we could. It wasn't too hard to build, but we never sold another one like it. Every once in a while I would get a letter or a phone call from a customer telling me burglars had broken into their house but couldn't get into their Kingsbery Gun Safe. It wasn't unusual for several of their neighbors to buy a safe.

Ironically the owner of the safe company in San Antonio who didn't have time to talk to me called me two years later. He wanted an exclusive dealership for South Texas. I didn't mention how he had treated me two years before but I thanked him and told him I wasn't interested. He called me a year later still wanting the dealership. We didn't have any competition for five years and even then people told me we made the best safes. One of the first models we built in 1971 was called the Fort

Knox. In 1981 I received a letter from a lawyer representing the Fort Knox Safe Company. The letter stated that I was illegally using the name Fort Knox in my advertisements and brochures and if I didn't immediately cease using Fort Knox he would sue Kingsbery Manufacturing on behalf of his client.

    I wrote the lawyer thanking him for his letter and told him that my company or I had never been sued and that I didn't like litigation. For that reason, I wrote, I was not going to sue Fort Knox Safe Company for using the name of one of my safe models. I enclosed an advertisement from the American Rifleman Magazine featuring my Fort Knox safe that came out eight years before the Fort Knox Safe Company was founded. I told him not to worry, that I wasn't going to sue and that his client could continue using the name they copied from my company and wished them success in their venture. I never heard from that lawyer or the company again.

    I enjoyed owning Kingsbery Manufacturing, particularly for all the great employees we had over the years. Phil Alcozer and his wife, Mary Jane, ran the company when I was away. Frank Garza and Pete Teran both worked there for 28 years before retiring. They could make a squeeze chute blindfolded. Ceferino Gonzales, Lupe Rodriguez and Raul Hernandez were also longtime employees who contributed to the success of the company.

    In 1997, just before I turned 75 years of age, I sold the corporation on the condition that the new owners keep all my employees and maintain the same top quality that Kingsbery Manufacturing was known for.

HORSE TALES

Our oldest child, Ann, loved horses. As a toddler she would pet our ranch horses on their noses as they came up to the fence around our yard. Ann loved to sit in front of me when I was riding close to the house. One day I rode by the house on my way to a pasture down the road when Ann, about three years old at the time, came running out wanting to ride. I was on Old Sorrel, a top cow horse I had owned for years. I had never ridden him with Ann but he was fairly gentle so I picked her up and put her in the saddle in front of me. Ann held on to the saddlehorn and started laughing and singing as we rode down the road.

When we came to the gate that went into the pasture I stepped off Old Sorrel, leaving Ann sitting in the saddle and holding on to the saddlehorn. I had a good grip on the reins as I started to open the gate. Ann made a sudden movement in the saddle, surprising Old Sorrel. If a big panther had jumped on his back it couldn't have scared him any more. He bellowed and made a sudden lunge by me. I managed to grab Ann by her left leg as Old Sorrel started pitching high and hard in a circle, while I held the reins in a death grip. Ann was hanging on to the saddlehorn, but I pulled her off, holding her by the one leg high enough so her head didn't hit the ground as she swung off.

As soon as I pulled Ann off Old Sorrel he quit pitching. I put Ann down and she cried a little but I bragged on what a good ride she made and she started laughing. I explained to her that Old Sorrel wasn't used to two people riding him and when I got off he got scared when she moved. Ann ran back to the house to tell her mother what happened. I got on Old Sorrel and finished my work, glad nothing worse had happened.

⬅——————➡

One year I was pasturing yearling cattle on irrigated oat fields near La Pryor. I had finished checking the yearlings and was on my way back to Crystal City pulling a small two-horse open-top trailer behind my pickup. My horse was a registered Quarter Horse named Leo Bar None that I had bought for $3,500 as a yearling from Bud Warren in Perry, Oklahoma. He was one of the best-bred colts in the country, by Mr. Bar None and out of a top Leo mare, and I was really being careful with him.

When I loaded him for the ride home I tied him with the hackamore reins running under the top board, not worrying about him jumping out of the open-top trailer. I was driving a Chevrolet pickup that had one-by-six boards running long ways in the bed with two-inch metal strips holding the boards in place.

Just as we were coming into Crystal City a big oilfield truck started to pass us. When the cab of the big truck was even with my trailer the truck driver saw a car coming toward him. He didn't think he could make it around my truck so he took his foot of the accelerator to start slowing down. When he did that the truck engine backfired like a cannon going off. It actually fired twice but once was enough. On the first backfire Leo jumped forward over the front of the trailer. Since the hackamore reins were tied under the top board of the trailer, he turned a flip and his back legs landed in the bed of the pickup. Leo's belly was on the tailgate and his front feet were hanging down between the pickup and the trailer.

I saw the whole thing in my rearview mirror and immediately pulled off the highway and stopped. The men in the truck had seen Leo jump out of the trailer so they stopped behind us. I jumped out and ran to the back of the pickup to check on Leo. I almost fainted when I saw that one of his hind legs had broken through one of the boards in the bed of the pickup and was sticking through the hole past his hock. I immediately grabbed his ear and twisted it hard—that will stop a horse from struggling. I knew that if I could keep Leo from flouncing around trying to get his leg out I might prevent a permanent injury to his leg.

Three men jumped out of the truck and came running to help. Travis King, a friend who happened to be driving by, also stopped to help. Travis quickly untied the reins from the trailer and held Leo's head with a tight twist of his ear. I took my rope off the saddle, doubled it over and slipped it under Leo's flank. By then another man had stopped to help, so we had a real crowd. I told the new arrival and the three men from the truck to get up in the bed of the pickup, two on each side of Leo, and grab the rope that was under his flanks. I unlatched the tailgate so it would drop down as Leo moved forward. I crawled under the pickup and held the leg that was sticking through the hole in the bed. I hollered for the four men to pull Leo up by the rope.

As they raised Leo I pushed his leg carefully through the hole. Just as his foot cleared the hole the tailgate came down and Leo hopped down between the pickup and the trailer. His leg was skinned a little but he was otherwise unhurt. I breathed a deep sigh of relief, thanked everyone and loaded Leo back in the trailer. The truck driver really apologized but I told him he couldn't help what had happened and thanked him for stopping to help.

Later that year I put Leo in race training and he won his first race. I sold him right after the race for $9,500. I still have that trailer but now have pipe bows over the top to keep anything from jumping out.

One of my best cowhorses was Bandy King. I bought him as a yearling in 1957 and used him for many years on the ranch and for calf and steer roping. Evelyn rode him in barrel races and our children rode him in 4-H rodeos and on trail rides while he was still a stud. I bred lots of mares to Bandy King before I gelded him when he was seven years old. In 1973, when he got too old to work cattle I gave him to L. B. Moss in Cleburne, Texas. He had been the pastor of the First Baptist Church in Crystal City from 1962 to 1967 and always admired Bandy King. L.B. took good care of him until Bandy King died of old age several years later.

In 1966 I decided to expand my horse operation and bought an irrigated farm south of Crystal City to establish Kingsbery Stud Farm and Training Stables. I designed and built a twenty-stall barn completely out of railroad cross ties. Pete Simpson and his family moved to the farm and started training horses for other people. I started acquiring top Thoroughbred and Quarter Horse stallions and developed an extensive breeding program. I had bought several brood mares from Jap Holman, who had raised good horses for years near Sonora, Texas.

Some of my Quarter Horse stallions were Pleasure Bar by Pleasure Fund; Double Q. Lee by Leo; and Top Cause by Rebel Cause. The Thoroughbred stallions were Sonora Chief by Raffles; Rare Rice by Olympia; Kings Comet by Bold Lark and Native Native by Native Dancer, one of the most famous Thoroughbreds of all time. Native Native was out of Mill House, a stakes-winning mare by Basis.

I saw Native Native advertised for sale in 1970. He had been nominated for the Triple Crown but had chipped a bone in his knee and never got to race. Jacque Goff owned Native Native and she was standing him at her Blue Hen Farm near Ocala, Florida. I called Jacque and in October of 1970 I flew to Ocala to see him. Jacque and I walked out in Native Native's paddock and he trotted up to me and let me pat him on his head. He was, without a doubt, the best looking horse I had ever seen. I knew I had to have him. Jacque wanted more money than I was willing to pay but after lots of phone calls I bought Native Native for fifteen of my top Beefmaster heifers and a lot of cash.

I advertised him in the December, 1970 issues of **Thoroughbred Record**, **The Blood Horse** and **The Quarter Horse Journal**. Native Native arrived at the farm on the tenth of January and by the tenth of May he had bred enough mares to pay for himself. There were very few sons of Native Dancer in Texas and lots of people wanted some of that famous bloodline. The next year I bred mares to him from as far away as California, North Dakota and even Mexico City.

Native Native's first Thoroughbred foal to race, Native Ground, won the five-and-a-half-furlong Jasper Purse at James Ellis park in Henderson, Kentucky, by five lengths over ten other good two-year-olds. His first Quarter Horse foal to race, My Native, won a four-furlong race at Ruidoso Downs in New Mexico by four lengths over nine other two-year-olds. Native Native was booked full each year I owned him. By 1976 his offspring had such good racing records that several people wanted to buy him so I sold him that summer.

In commenting on Native Native's confirmation, one good horseman told me, "If I had made him out of clay and had a handful left I don't know where I would put it." A rancher from Waco brought a mare to breed and went back bragging about Native Native, saying he couldn't find any faults in his conformation. A friend of his, an old horseman who raised Quarter Horses and had judged lots of Quarter Horse shows, told him he had never seen a horse he couldn't find some fault in his conformation. When the rancher returned to pick up his mare the old horseman came with him. We all walked out in the roping arena where Native Native was standing and looking at some mares out in the pastures. As the horseman walked around Native Native several times the rancher who had brought him said, "Now don't make Jack mad."

I told the rancher that I wouldn't get mad, I wanted to know what he would change. After a few minutes the old horseman came over to me and said he had finally found one fault with Native Native. He said, "The son of a gun is in your pen and not mine!"

Pete Simpson moved to Batesville and Sammy Gonzalez ran the horse farm and trained horses for me. We continued breeding and training horses until 1984. Oil was selling for $30 a barrel and everyone and their brother-in-law wanted to get in the horse business. I figured it was time to get out so I sold all my horses except the cowhorses that I used on the ranch. A couple of years later oil prices dropped to $10 and the bottom fell out of the horse business.

I still use the horse farm in my cattle operation, but it makes me sad to see all those empty stalls in that good horse barn.

COUGARS, COYOTES AND MOTHER HENS

Prince Wood and I started ranching south of Batesville, Texas, on part of the George West Ranch. We were both single and lived in a big house on the ranch, doing our own cooking. It was eighteen miles to the closest town and we didn't have a telephone. We decided we needed some hens for fresh eggs so we bought a dozen White Leghorn layers. We put them out at the ranch and they started laying eggs right away, but within a few months the coyotes had caught every hen.

Someone told us to get game hens because they were so fast and agile that the coyotes couldn't catch them. We bought ten game hens and a game rooster. The game hens lasted longer than the Leghorns but we were still losing about one a month to the coyotes. After six months we were down to two or three hens and the rooster.

One game hen had nested under the barn and hatched twelve baby chicks. When the chicks were about a week old I was at the barn a little after daylight to feed the horses. The mother hen and her chicks were scavenging for food about thirty yards from the barn near the brush. I looked out at the hens and chicks just in time to see a big coyote slinking toward the hen. Before I could holler the coyote made a dash for the hen. She flew up and toward the coyote to protect her babies. The coyote just snatched her out of the air, wheeled and ran back into the brush with the mother hen still flopping in its mouth.

I had to catch the chicks, build a stout chicken coop for them and be their mother until they were big enough to make it on their own. We kept buying our eggs in town.

◆―――――▶

There have always been mountain lions, also called cougars or panthers, in South Texas, but you seldom see one. Their main source of food is deer, but they will also eat sheep and goats. Their favorite meal is a young colt. If there is a colt in the area it will be the first thing a cougar will kill. In my forty-eight eight years of ranching I have had only two encounters with panthers.

Prince and I had a remuda of about ten cow horses in a pasture north of the headquarters. One morning we couldn't find the horses and after looking for an hour we found where they had run through a five-strand barbed wire fence, breaking all the wires. We followed their tracks through the neighboring field and found where they had run through another fence, breaking all five wires on that fence. We finally found the horses in the next pasture and drove them back to the corrals. Five of the horses were cut from the barbed wire but none of the injuries were serious.

I went back to the horse pasture to find out what had spooked the horses so badly. Close to a tank in the pasture I found a big buck deer that had been killed during the night. Part of the deer had been eaten and it was partially covered with brush. I could tell by the tracks around the carcass that a big panther had killed the deer. By reading the tracks I found where the panther had crouched on a mesquite limb over a trail leading to the tank. The panther had jumped on the buck as it walked under the limb on its way to drink out of the tank.

It was hard to believe the fury of the life and death struggle between the panther and the buck. At least half an acre of brush and tree limbs had been torn up in the battle. The horses were sleeping nearby and the sudden wild commotion scared them so much that they ran through two fences before stopping.

I had been told that a panther returns to its kill the following night. Prince and I wanted to get the government trapper, Bob Bragg, who lived in Batesville to come to the ranch and set some traps to catch the panther. We wanted to get rid of the panther because we were planning to raise colts and didn't want to lose any. We didn't have a phone and were busy that day working cattle so we didn't have a chance to call Bob. We checked the next morning and, sure enough, the panther had come back and eaten more of the deer.

About a year later I was riding down a dusty cattle trail in the same pasture when I noticed panther tracks heading toward the tank. I followed it to the five-foot net wire fence near the tank and saw where the panther had jumped the fence and landed about ten feet on the other side. I tracked him farther, past where the buck had been killed the year before. His tracks showed that he walked among the few deer bones that were left. I'm sure it was the same panther.

My second encounter came twenty-five years later at my ranch west of Crystal City. Our youngest son, Bruce, and I were driving on the ranch when a big panther ran across the road in front of us. That was the first panther I had ever seen out in the open. A few days later a friend was hunting deer on the ranch, sitting in a deer stand. He saw a panther and shot at it. He thought he hit it, but couldn't find the panther so he came to our house to tell us about it. Bruce and I went back to the ranch with him and found the panther's tracks but no blood.

About a week later we rounded up cattle in that same pasture and penned them in the corrals by the highway that ran along one side of the ranch. I had sold two truckloads of Beefmaster cows and calves to a rancher in Florida and the trucks were coming to pick them up the next day. My cattle were very gentle and the corral was made of railroad cross ties and heavy net wire.

Gene Allen from Carrizo Springs was hauling the cattle to Florida and the next day his trucks arrived at my corrals at the same time I did. To everyone's surprise the cattle were nowhere to be seen and one whole side of the corral was lying flat on the ground. We rushed to our farm south of town to get our horses so we could pen the cattle again. Usually my cattle are easy to pen but that day they were really leery about going back into the corral. We finally got them penned and on their way to Florida.

I'm sure that the same panther I had seen before came by the corrals or got into the corrals and that's what scared the cattle so badly. I saw him again about three days later but never saw the panther again after that.

MESQUITE CORRAL COWBOYS

Bob Woodward, Prince Wood and I were running about 500 cows on a ranch south of Sabinal. Prince and I lived on the ranch and did all the cattle work ourselves. When it came time to work the calves we would pen about fifty or sixty pairs in a large corral. One of us would ride into the herd, rope a calf, dally the rope around the saddle horn and drag the calf out of the herd. The other person would throw and hold the calf while the roper got off his horse to castrate, vaccinate and earmark the calves.

After one of us had roped about ten calves we would switch and the other person would get on his horse and rope the calves. We would continue this procedure until we had worked all the calves. One day we had a bunch of calves that needed to be worked. We had gathered the cows and calves out of a large pasture the day before and had them in a holding trap next to the corrals. The corrals at the ranch were the "wood filler" type that had been built in South Texas for more than a hundred years. They were made by putting two rows of mesquite posts about twenty inches apart and stacking smaller mesquite poles horizontally between them to about five feet high. Wire is wrapped around the vertical posts to keep them from spreading as they are filled with the mesquite poles. This type of corral will take lots of abuse and last for many years.

Since it was summertime we penned the cattle just after daylight so we could get part of the calves worked before the blazing sun started working on us. There were about 75 crossbred cows with calves sired by Angus bulls. There had been lots of green grass that spring and the calves were fat. Most of them weighed about 300 pounds and it sure strained a person to throw and hold one down while the other worked on it. Just as we closed the corral gate three of our calf roping buddies, Marvin Angermiller and Ross Martin from Uvalde, and Hose O'Bryant from Utopia drove up.

Our three friends walked up to the corral and looked over the fence at the calves we were about to work. Marvin Angermiller said that the calves are just what they needed to practice tying because the rodeos were using bigger calves in the calf roping events. Prince and I told them we were just getting ready to work the calves and we would let them throw and tie the calves for practice. The three had just glanced over the corral fence and hadn't seen all the calves. I told them they could do it only if they would throw and tie all the calves in the corral. They agreed and got their pigging strings out of their pickup. We already had our knives sharpened, vaccine ready and a can of screwworm repellant with the swab brush ready to go.

    Prince and I were all smiles as we rode into the corral with our ropes ready. Since we both loved to rope this would be great, just rope and drag the calves up to our friends and let them do the hard work. I'm sure our friends didn't realize how big and rank those half-Angus calves were, or how many there were when they agreed to throw and tie all of them. For awhile our friends were getting a big kick out of throwing the calves, tying them with their pigging strings and throwing up their hands like in a rodeo. It wasn't long until they began to realize that the calves were bigger and stronger than they had thought. By the time they would finish working one, untying it and letting it up Prince or I would be dragging up another calf for them to throw.

    By the time we had worked about half the calves the sun was blazing down on us and it was sure getting hot. All three helpers were covered with sweat and dirt. One had a big hole torn in his Levis and another had a shirt pocket torn completely off. The calf ropers said they were really getting thirsty and needed a drink. I told them that Prince and I weren't thirsty but we got off our horses and all went in the house to drink iced tea and water. We sat around the kitchen table for a good while telling stories about calf roping.

    After we had been in the house about thirty minutes, Ross Martin looked at his watch and said, "We need to go, I've got to meet Mr. Baker at the ranch in a little while." Hose O'Bryant said he sure needed to get back to Utopia. Then Marvin remembered some reason he needed to be back in Uvalde. I said, "Wait a minute. You all agreed that if we let you throw and tie the calves you would tie them all." They did some griping but we all went back to the corral and finished working all the calves. Prince and I thanked them and they said they were sure going to take a better look at all the calves before they made another deal like that.

A few months later Prince and I penned a three-year-old Braford bull in those same corrals. He had screwworms in his sheath and needed doctoring. We had cut him out of the cowherd and he was really fighting our horses by the time we got him to the corral. The year before we had bought 45 Braford bulls from Mr. Claude McCann of Victoria, Texas. This bull was the only one in the bunch that was spooky. Even though he had been there over a year he didn't seem to be getting any gentler. He had fairly long horns that we had tipped so they wouldn't be too sharp.

We were going to have to throw the bull down to do a good job of killing the screwworms. Prince was a better heeler so I usually roped the head and he would rope the heels. As we rode in to the corral the bull made a run at us. As my horse dodged I threw my loop at his head but missed. As the bull went by Prince roped both hind legs, wheeled his horse and started dragging him. I jumped off my horse, grabbed the bull's tail and tried to jerk him down. This is a common practice and is usually easy to do if the animal isn't too big and stout.

I jerked his tail at a right angle to the direction Prince was pulling and got him down twice. He was so stout I couldn't hold him down. Prince was looking back at me and didn't realize he was heading straight into a corner of the corral. By the time Prince looked around and saw he was in a corner it was too late to wheel his horse around to keep the rope tight. I was hanging on to the tail and pulling with all my might to throw the bull one more time. When Prince's horse reached the corner and stopped the bull backed up a little bit and just stepped out of the now slack loop.

There I was, on foot in a big corral holding a very mad bull by the tail. I wheeled and ran for the closest fence as fast as I could with the bull right behind me. Just as I got to the fence he caught up with me, lowered his head, hooked my behind and threw me completely over the fence. Luckily I landed in deep sand on the other side and wasn't hurt at all. The bull's right horn had hit my billfold in my hip pocket and that kept me from getting a badly bruised hip.

I got back on my horse and we roped the bull and stretched him out on the ground to doctor. Prince said he sure wish he'd had a camera to take my picture as I sailed over the five-foot fence.

WILD STEERS IN A GARAGE

Clyde Watkins called one day, wanting me to bring my two good leopard cowdogs, Mitzi and Lady, to help catch three steers on the W.E. Lee Ranch south of Uvalde. Clyde ran lots of cattle in South Texas and had some top cowboys working for him so I knew the steers had to really be wild. He said the four-year-old Hereford steers had gotten away from his cowboys the last two years. Clyde said the steers stayed in the heavy brush where the Leona River ran through the eastern side of the ranch. At that time the river had water in it and was pretty wide and deep.

The next day, before daylight, I loaded my good brush cow horse, Old Sorrel, in my trailer and called for Mitzi and Lady. Knowing they were going to work cattle, the two dogs eagerly jumped in the back of the pickup and we headed out for the ranch. Clyde drove up just as I arrived at the east entrance to the ranch. His brother, Shelly, Jack Miller and two other cowboys whose names I can't remember were with him. We unloaded our horses and I motioned for Mitzi and Lady to start tracking. They took off into the heavy brush looking for cattle.

The Leona River was about a quarter of a mile from the east fence of the ranch and we started riding north from the middle of the pasture. Soon we heard the dogs baying which meant they had found the steers. We took off at a run in the direction of the barking. After we had run a ways we stopped to listen. I told Clyde I thought the steers were coming back toward us but Clyde thought they were headed north. Clyde and his cowboys took off in that direction but I sat there on my horse to get a better idea as to which way the steers were traveling. In a few minutes I could tell that the steers were heading back south. There was a narrow utility right-of-way running between the river and the east fence all the way through the ranch. I had stopped my horse on the edge of this right-of-way and could tell by the barking of Mitzi and Lady that the steers were headed right toward me.

I grabbed my rope and had my loop built, hoping the steers would come by close to me. In a minute I saw the three big steers coming down the right-of-way toward me with Lady and Mitzi right on their tails. I was sort of hidden behind a thick bush. As they went past me I took off after them and roped the last steer before he had time to duck off into the brush. I followed him into the heavy brush until I came to a big mesquite tree where I could neck him. About that time the other cowboys rode up and helped me tie the steer to the tree. They all commented on how brushy it was and they didn't know how I could rope him in such thick brush. I didn't tell them that I roped the steer in the right-of-way.

Mitzi and Lady, as usual, stayed after the other two steers. By the time we necked my steer the dogs were out of hearing. We all sat on our horses and listened but couldn't hear them barking. After awhile Clyde said he didn't believe the dogs were still with the two steers. I told him that Mitzi and Lady were with the steers somewhere in the pasture. Even though we were in a big pasture the wind wasn't blowing so we should have been able to hear the dogs barking a long way off.

After a good while Clyde wanted to ride on south and look for the steers. I kept insisting that we wait a little longer. If the steers had swum the river and gone west in the more open country they would soon head back to this side of the river where they preferred the thick brush. I knew my cowdogs were still with the steers. In a little while Jack Miller said he thought he heard a dog bark and pointed southwest. We all rode in that direction and came to the river where we could hear a faint barking. The dogs were across the river and they weren't moving, which meant the steers had stopped.

The river was too wide and deep to swim our horses across, so we loaded them in our trailers and drove the 14 miles to the west entrance of the ranch. As we got close to the river we stopped to listen for the dogs. We could hear them baying the steers near an old abandoned house. When we drove up a little closer we could see Mitzi and Lady barking into an old garage. The two steers were hiding in the garage! Since the garage was facing west we couldn't hear the dogs when we were on the other side of the river. One of them had probably moved outside the garage and that was when Jack Miller first heard the barking. By then Mitzi and Lady had held the steers in the garage for about two hours.

We all got on our horses and made a big circle to come up to the garage from the rear so the steers would take off to the west where the country was fairly open. We didn't want to run the steers too far before we roped them and didn't want them to cross the river again. When the steers saw us they took off and we quickly roped them. As we loaded the steers in the trailer Clyde really bragged on Mitzi and Lady and how they stayed with the two steers across the river and into the garage. Later I gave Clyde one of Mitzi's puppies. He named her Queenie and used her for many years to catch wild cattle.

MAURICE AND THE RATTLESNAKE

Of all the stories I told my children about my younger days, for some reason this was the one they always liked the best.

My younger brother, Tom, and I often went fishing. We had several large tanks on the ranch and they all had fish in them. One tank, close to the house, had lots of crawfish in it and when we were little we liked to catch them. We would use a small tree limb with a line and chunk of fat meat tied to the end of the line. We would lower the meat into the water and a crawfish would catch it with its pincher-like claw and hang on long enough to be pulled out of the water. We would put them in a bucket and if we caught enough of them Mother would fry the tails for us to eat.

When Tom and I were twelve and thirteen years old we decided to go fishing on Home Creek. It was a big creek about a mile from the south end of our ranch. We planned an overnight fishing trip and Mother helped us put our supplies together for fixing our meals. We got some blankets to sleep on and an axe to cut wood so we could make a fire for cooking. Our six-year-old cousin, Maurice Kingsbery, lived about a mile east of us on another part of the Kingsbery Ranch. He really wanted to go with us so we told him he could.

We went down to the little tank below the house and caught a bucket of crawfish and small perch for bait. We put the harness on two of our workhorses and hitched them to a wagon. We loaded our supplies in the wagon and headed to Home Creek. We got there late in the afternoon and the first thing we did was set out a bunch of throw lines from the creek bank. We baited the lines with crawfish and perch and tied them to tree limbs so we could see the limbs moving if there was a fish on the line.

While Tom and Maurice were finishing setting out the lines I unhitched the horses from the wagon, staked them out and started unloading the wagon. I took the axe and walked down the creek to chop some wood for our campfire. As I was walking I came up on a big rattlesnake. Luckily, I saw it before I got too close and killed the rattler with the axe. As I was carrying the firewood back to camp I dragged the rattlesnake along to show him to Tom and Maurice.

They were still down by the creek when I got back to camp so I coiled the dead snake and put a tow sack over it. We had a kerosene lantern that I lit and hung in a tree right above the sack covering the rattlesnake. While we were cooking supper by the light of the lantern I told Maurice to hand me the tow sack so I could put some trash in it. When he picked up the sack he saw the big rattler and jumped back, scared to death. Then he saw it was dead and he wasn't too upset at me, even laughing about it as we were eating supper. But Maurice did say he would rather sleep in the wagon than on the ground with Tom and me.

It was a mean thing to do to my little cousin, but I guess about the norm for ranch kids that age. We caught ten or twelve good size catfish that night, but didn't see anymore rattlesnakes.

Many years later I married Evelyn Bruce and we lived on a ranch in South Texas. We had four children: Ann, Bob, Kay and Bruce. Our parents, Charles and Leona Bruce and Howard and Mabel Kingsbery, lived in Santa Anna. Every time we went to visit them we crossed a bridge over Home Creek and our kids would say, "Daddy, tell us about 'Maweese' and the rattlesnake." For some reason this particular story just fascinated our children. I always had to tell the story as we drove into Santa Anna.

One summer our six-year-old grandson, Clay Kingsbery, was visiting us from Washington State. We were on our way to Santa Anna and were about to cross the same bridge when I mentioned something about Home Creek to Evelyn. From the back seat little Clay piped up and said, "Grandaddy, tell me about Maweese and the rattlesnake." Evelyn and I both laughed, remembering how our children always insisted on hearing the story as we crossed the bridge.

I mentioned the incident to my son, Bob, a few days later and he said he wasn't surprised. He had told Clay about Maurice and the rattlesnake when he was about two years old. Then he had to tell Clay that same story every night for the next two years! The story just goes on and on.

←——————————→

Rattlesnakes are everywhere in South Texas and you always have to be on the lookout for them. Here are some of the more memorable rattlesnake encounters I have had during my 49 years of ranching in South Texas.

During the terrible drought of the 1950s I was burning prickly pear cactus to feed 500 cows on a ranch south of Batesville, Texas. We used a portable butane burner to singe the thorns off the cactus leaves so our cows could eat them. Prickly pear with the thorns burned off makes a pretty good substitute for grass during a South Texas drought. We had been doing this every day for about a year and I had a "mojado" (illegal alien) from Mexico named Beto helping me. He was the best worker and fastest pear burner I had ever seen. I gave him an extra bonus each week and told him it was for doing such a good job and burning so much pear. He would send a money order to his family in Mexico every week.

Burning pear was no fun, besides the incredible heat and carrying a heavy burner on your back all day, you had to watch out for rattlesnakes. Almost every big pear bush had a cactus rat's nest in it, and most of the rat nests had rattlesnakes in them. When we burned pear the rats' nests would catch on fire and the rattlers would come slithering out. The rattlesnakes were just trying to escape and weren't really dangerous unless you stepped on one.

To avoid the heat of the South Texas sun we would start burning pear about four o'clock in the morning and burn until around noon. One morning Beto and I were standing about fifteen feet apart, our burners going full blast, when I happened to glance over toward Beto and saw the rattlesnake. It was huge, at least six feet long and moving fast away from the burning cactus right between Betos feet. I didn't think Beto was in any danger as long as he didn't move so I didn't say anything.

Beto was wearing homemade huaraches—tire-tread sandals with two leather straps, leaving most of the top of his foot bare. Beto hadn't seen the snake and everything would have been fine, except the snake changed direction and started crawling over one of Beto's feet. Beto looked down and saw the huge rattlesnake and turned as white as a suntanned mojado could. He quickly slipped the butane burner off his back, dropped it to the ground and ran backward about twenty feet.

I killed the big rattlesnake with my pear burner and turned his burner off. I kept on burning pear and after a few minutes I motioned for Beto to come back and start burning pear. He just shook his head. After awhile I turned my burner off and told Beto we still had lots of pear to burn. All he said was,

"Voy a Mexico" (I'm going to Mexico). No amount of talking could get him to change his mind so I told him I would have to go into Batesville to get him the money for the two or three days he had worked that week.

All he said was "No quiero dinero. Voy a Mexico" (I don't want any money. I'm going to Mexico). I took Beto back to the house where he was staying so he could pack his stuff. I thanked him for his hard work and he headed out for Mexico. I had his address in Mexico so the next day I sent him his money plus a bonus. Unfortunately I had to burn pear until dark that day.

One fall day I was helping round up steers on the Smyth Ranch southwest of Uvalde, Texas. There were six of us cowboys riding single file down a cattle trail in the heavy brush. The trail had washed down about a foot deep as we neared a little creek. I was third in line when all of a sudden a big rattlesnake reared up and struck the toe fender on my stirrup. The force of the strike felt just like a person had thrown a fist-sized rock real hard. The blow spooked my horse and, as he was scrambling sideways to get out of the deep trail, he fell to his knees. I didn't get off because the rattlesnake was too close. My horse got back on his feet and we killed the rattler, which was over six feet long.

The rattlesnake left a lot of yellow venom on the toe fender. I was wearing boots and leggings made of heavy leather for cowboying in the thick South Texas brush. The rattlesnake's fangs probably couldn't have penetrated even if it had struck higher. I figured the rattler had been coiled up by the trail waiting for a rat or rabbit to come along. The first horse in line made him mad and the second horse made him even madder. By the time the third horse, the one I was riding, passed him he was mad enough to strike. From that day on I made sure I was always the first or second in line.

The drought ended in 1957 and South Texas had good rains for several years. By 1959 there were more rattlesnakes than even the old-timers could remember. The abundance of grass and weeds caused the rat and rabbit population to explode. They are the main diet of rattlesnakes. We lived on a ranch west of Crystal City and our four children must of thought their first names were "Watch out for rattlesnakes" and "Cuidado para las viboras." That's what my wife Evelyn and Concha, the Mexican girl who lived with us, told them every time they left the house. We killed 13 rattlesnakes around the house and barn that year.

One night after dark I drove in from another ranch with two horses in the trailer. I unsaddled the horses and took them into the corral with a bucket of feed. I poured some of the feed in one trough and, as I started toward the second trough, I heard a rattlesnake start rattling right in front of me. It was pitch dark and I couldn't see anything so I just stopped and stood perfectly still. All of a sudden another rattlesnake started rattling right behind me. I was afraid to move because there could have been a third rattlesnake nearby. The two kept rattling so I started calling for Evelyn as loud as I could.

She finally heard me and came from the house to see what I wanted. I told her to bring the shotgun and a flashlight. She shined the light all around and located the two snakes. They were far enough from me that I could reach for the gun. With Evelyn holding the light I killed both rattlers and they were each about four feet long. They never quit rattling all the time I was waiting for Evelyn to bring the shotgun.

During that same time we had several cats around the house and they didn't like rattlesnakes either. The cats would try to chase the rattlers away from the house, causing them to coil up and rattle. That alerted us that a snake was near the house. We usually kept a 22-caliber pistol loaded with rat shot and a 410-gauge shotgun in the house to kill snakes.

One morning when I was gone Evelyn heard a snake rattling under the house, which was raised about a foot and a half above the ground with no siding around the bottom. She looked under the house with a flashlight and saw a big snake coiled up and rattling. She went back into the house for a gun but couldn't find the pistol or shells for the shotgun. With four little kids in the house, Evelyn wanted to make sure she killed the rattlesnake. She got my 270-caliber deer rifle with a scope and got down on her hands and knees with the rifle and flashlight and maneuvered to a position where she could shoot the snake without hitting any gas or water pipes. In that kneeling position the rim of the scope was just above her eye. She lined up the crosshairs of the scope on the snake and pulled the trigger. The blast killed the snake but the recoil of the gun caused the scope to make a half-moon shaped cut above her eye. I was proud of Evelyn for killing the snake and was glad the cut healed without leaving a scar.

Fortunately no one in our family was ever bitten by a rattlesnake but several of my cowdogs, cattle and horses were bitten. Young calves and colts sometimes die from rattlesnake bites. In their curiosity about a coiled rattler they put their noses close to smell it and the snake strikes. The venom causes the head to swell so badly the little animal is unable to suck milk and starves to death. People in this area are always on the alert for rattlesnakes because their deadly bite is an ever-present threat.

LOVE, WAR AND CATTLE PRICES

A West Texas rancher I knew was involved in a bitter divorce case. One major issue was the value of 300 head of cattle jointly owned by the couple. The lawyer for the rancher's wife asked me to appraise the cattle. I accepted and we all met at the ranch headquarters with the rancher and his lawyer, who had brought their own appraiser.

I had been on the ranch several times and knew the cattle well. In fact, I had bought several bulls from the ranch the year before. As we were drinking coffee before going to look at the cattle I asked the rancher if all the cattle were certified Beefmasters. The rancher spoke up loud and clear, "They are all crossbred cattle." I was surprised to hear him say that since I was a director of the breed association, Beefmaster Breeders Universal, at the time and knew the cattle were certified Beefmasters. The rancher's lawyer also stressed that the cattle were crossbred cattle and should be valued as such.

West Texas had been under a drought for two years and cattle prices were really down. Good crossbred cows were bringing about two hundred dollars while certified Beefmaster cows were bringing four hundred to five hundred dollars. I wasn't surprised the rancher wanted the cows appraised as crossbred cattle; it would save him a lot of money when it came time to buy out his wife.

I didn't say anything as we drove through several pastures looking at the cattle. The other appraiser and I were given sheets of paper with the number of cattle in each pasture and their age. I noticed that every one had a BBU ID number branded on its side as well as the ranch holding brand. After inspecting the cattle we drove back to the ranch headquarters. We all sat around the dining room table as the other appraiser and I started figuring up the value of the cattle. The rancher's lawyer commented again that these were crossbred cattle and we should both appraise them accordingly.

We each wrote down on a piece of paper the market value that we placed on the cows. The lawyer for the rancher read out loud the value the appraiser for the rancher had written--two hundred dollars each. The rancher said that was about right. Then the lawyer read my appraised value: Four hundred dollars each.

I thought the rancher would fall out of his chair when he heard the amount. He really started putting me down, saying I didn't know what I was talking about. I didn't say anything until he finished. Then I told his lawyer to read what I had written at the bottom of the page.

He picked up the paper again and read out loud, "This is my appraised value of the 300 cows and also my bid for the cows. I hope you will sell them to me."

The rancher and his lawyer just sat there in shocked silence for several minutes. Finally the wife's lawyer and I thanked the group, got up and left. I not only lost the rancher's friendship, he didn't sell me his cattle either.

Another time two South Texas ranching partners had a huge falling out. They ran about 160 good crossbred cows and had just shipped all the calves to market. It was a fifty-fifty partnership so the profit from the calves was split two ways. Both men wanted to keep their share of the cows since they were top quality and would be hard to replace. Due to the ugliness of the breakup neither rancher trusted the other to divide the herd fairly. They argued a lot and almost got in a fistfight while trying to figure out how to divide the cattle fairly.

Since they were both good friends of mine they asked me to divide the cattle for them. The cows were pretty uniform and all about the same age. My first suggestion was to run the cows through the chute and cut one to the right and the next one to left. At first they thought that was a good way to divide the cattle. Then, after thinking about it for a while, one rancher thought his pen of cows might not be as good as the other pen so they turned that idea down. My second suggestion was to do the same thing then flip a coin. The winner would get his pick of the two pens. That wasn't accepted because one rancher thought he might not get first pick and his ex-partner would get the best cattle.

I told them that they were sure being "picky" in worrying about the other person getting the best cattle. I told them that the cattle were so uniform that each one would be getting the same quality cattle. Neither rancher would budge. It had been a bitter breakup so I had to come up with a method of dividing the cattle worthy of Solomon.

The cattle were real gentle, so I made the third suggestion for dividing the cattle that would be slow but fair. I wanted one of the ranchers to start by cutting out two cows as near the same quality as possible. Then the other rancher would pick one for himself and the other cow would belong to the man who had cut the cows out of the herd. Then the procedure would be reversed until all the cattle were divided. After thinking it over both ranchers agreed to the method. It took several hours to divide the cattle that way but both men were satisfied when they finished. They both thanked me and seemed to be in a good mood when I left.

THE KILLER BULL

When I turned 70 years old I began cutting back on my ranching operation and now run about 100 Beefmaster cows on our ranch west of Crystal City. My wife, Evelyn, and I do all our own ranch work. We have two good cowhorses, April and Dolly, that we ride when working cattle.

One Saturday afternoon in December of 1996 I got a call from Luis Contreras, Crystal City's chief of police. He told me a bull had just killed a man on a little farm on the edge of town and wanted me to come quickly and tell them what to do. I drove to the farm, where about a dozen people were gathered.

Chief Contreras told me the bull had seriously injured Manuel Matamoros on Thursday and that he was still in the hospital in San Antonio. Earlier that day the bull had attacked and killed its owner, Victor Bonilla, who had raised it from a calf. Chief Contreras said the bull killed Victor around noon and his son found him about one o'clock. The bull, a four-year-old black Jersey with sharp horns, had trampled and dragged Victor and hooked most of his clothes off.

The bull was out in a small pasture with two yearling heifers. The "corral" was a small pen, fenced with four barbed wires with a flimsy shed in one corner. The bull was really mad—pawing the ground and shaking his head as if daring anyone to come out in the pasture. Chief Contreras thought it would be safer to shoot the bull right where it stood before it attacked someone else. I told the Chief I thought the best thing to do was rope the bull, load him in a trailer and haul him to the auction in Uvalde. The chief thought that would be too dangerous but I said I could do it if I could find Lupe Rodriguez to help me.

Lupe was a top cowboy who worked at my manufacturing company as a welder. I called Lupe's house on my cellular phone and he said he would be there in a few minutes. After he arrived I told all the spectators to move back from the fence and wait until Lupe and I came back with the horses.

We went to my house to get our horses. I told Evelyn what had happened and that Lupe and I were going to rope the bull and load it in the trailer. Evelyn was very disturbed about our using our good cowhorses to catch a dangerous bull that could disable one of them. She said our horses were worth twenty times as much as the bull and that we shouldn't take the risk of getting one of them gored. She wanted me to let the police chief shoot the bull in the pasture and haul it away.

I was confident that Lupe and I could catch and pen it. As we went out the door, Evelyn cautioned Lupe to be careful to keep the bull from goring her mare, April. Lupe and I caught and saddled the horses, loaded them in the trailer and headed back to the Bonilla place. As we were unloading the horses and tightening the saddle girths a young man walked up and said, "Mr Kingsbery, I sure hope you can catch that bull. He just killed my daddy."

The killer bull was still bellowing and pawing the earth but had come into the corral with the two heifers after the spectators moved back from the fence. I backed the trailer up to the corral and opened the tailgate. Lupe and I rode into the corral with our ropes ready. I was sure we both would have to rope the bull and drag him into the trailer. The bull charged as soon as we rode into the corral, trying to hook our horses. We managed to dodge the bull and started hollering and driving him and the heifers toward the trailer.

Just as they got near the trailer the two heifers ducked back and Lupe and I ran into the bull with our horses, pushed him into the trailer and slammed the tailgate. We put the bull in the front part of the trailer and loaded April and Dolly in the back. We took the bull to my farm below town to keep him until the cattle inspector could haul him to Uvalde the following Thursday. All the way to our farm the bull kept trying to hook the horses over the middle gate of the trailer, but we made it without any problems.

We put the bull in one of my strong pipe pens, made from steel pipe. As I walked along next to the pen the bull would follow me, bellowing and trying to hook me through the fence. The next day Evelyn drove to the farm to see the killer bull. When she pulled up in her car the bull began to bellow and paw the ground. He looked so mean that she wouldn't even get out of the car. The next day Robin Clark, the cattle inspector, hauled the bull to Uvalde to be sold for slaughter.

People around Crystal City made the comment that, although there were lots of young cowboys in the area, the police chief called a 74-year-old cowboy to catch the killer bull.

ALWAYS CLOSE THE GATE

My wife, Evelyn, is an excellent cowhand. She grew up on a ranch with two sisters and the three of them were their daddy's cowboys. They learned at an early age to always close the gates on the ranch after you went through them. That habit has stayed with Evelyn until this day. Lots of times we'd go through a gate at the ranch and I would say, "Leave it open, we're coming right back." But she would close it anyway.

One day we were going to round up the cattle at our ranch near Crystal City. Randy Grissom, a local rancher, had bought two young bulls from me and was going to pick them up at eleven o'clock that morning. Our cattle will usually follow my pickup to the corrals in anticipation of the cottonseed cake I always put out.

It was a dry year and our cattle would immediately head for the corrals when they saw my pickup coming through the pasture or I honked the horn. I thought we could pen the cattle that way and wouldn't have to haul our horses to the ranch to pen them. We keep our horses on our farm three miles south of Crystal City, about thirteen miles from the ranch.

When Evelyn and I got to the ranch that morning the cattle were on the opposite side of the pasture from the corrals. A strong wind had just blown up and the cattle were too far from the corrals to hear the pickup horn. Since Randy was coming to pick up the bulls in about two hours we hurried back to the farm to get our horses.

As we drove along the county road next to the ranch I saw some of my cows about a quarter of a mile from the road. I wanted to see if the two bulls were with those cows so Evelyn opened the pasture gate and I stopped the pickup in the middle of the gate and told her to get in because we would be back in a few minutes. She had that look on her face that said we should shut the gate but she got in and didn't say anything.

I had about a hundred head of cattle in that pasture so I honked the horn and all the cattle came to the pickup including the two bulls. The cattle would have followed us to the corrals but one of the creeks that ran through the ranch had water in it and we couldn't cross it in the pickup. I hurriedly put some cottonseed cake on the ground to hold them there, jumped in the pickup and headed back to the road. As we drove out through the open pasture gate Evelyn got ready to get out and close it. I told her that the cattle would stay where I had put out the cake. She said we should close it but I kept right on driving and sped toward the farm.

At the farm we quickly saddled two horses, loaded them in the trailer and headed back to the ranch. As we came up over a hill about three miles from the ranch I could see a big herd of cattle in the highway heading toward us. My heart skipped a beat and I hoped I would see cowboys driving the cattle as we came over the next hill. If there were cowboys with them then the cattle wouldn't be mine.

As we got closer we could see that the highway was full of Kingsbery Beefmaster cattle heading right toward us at a rapid pace. It hadn't take them long to eat the cake I put out and then follow the pickup right out the gate wanting more. Evelyn didn't say a word. We pulled over and unloaded her horse so she could turn the cattle back toward the ranch. I drove ahead of the herd while she drove the cattle about three miles to the ranch gate. It was slow going because the cattle wanted to stop and eat the new green grass and weeds that were growing in the right-of-way along the road.

I got on my horse when they got to the gate to turn them into the pasture. As the last cow went through I got off my horse and closed the gate, even though we were going to pen the cattle and come back to get the pickup. I had learned my lesson. By the time we got the cattle to the corrals Randy was waiting to pickup up his bulls. We got the cattle penned without any more trouble so he didn't have to wait too long. I hated to admit that I had made a mistake, but it helped that Evelyn never said, "I told you so."

Evelyn and I regularly drove up to Santa Anna, Texas to visit our families. During those trips in the late 1960s and early 1970s we enjoyed visiting with Ace Reid and his wife, Callie. The Reids had ranched near Burkburnett in North Texas for many years and often drove through Santa Anna on their way to Kerrville to visit their son, Ace Reid Jr., the famous cowboy cartoonist. Ace Sr. liked the looks of the town so when he sold his ranch he bought a house in Santa Anna and soon made friends with just about everybody there.

Ace was certainly a colorful individual and in a short time he and Evelyn's father, Charlie Bruce, became great friends. They were both in their seventies and were old-time cowboys. Ace always wore an old-fashioned wide-brim Stetson with a pointed crown and stuck his pants in his high-top cowboy boots. He could tell funny ranch stories all day long and it was easy to see where Ace Jr. got the background for his cartoons.

Ace Sr. would go with Charlie nearly every day as he drove to his different ranches to check on his cattle and horses. They joked and kidded each other all the time and were good companions. One time when we were visiting I went with Charlie and Ace to move some cattle. The Santa Fe railroad tracks went through one of Charlie's properties and we were going to drive some cows from one pasture to another across the fenced-off railroad tracks.

Charlie said we didn't need any horses since the cows would follow his pickup looking for cottonseed cake. Charlie drove into the pasture and honked the horn. The cows came to the pickup and we drove slowly up the lane to the railroad tracks. Charlie told Ace to get out and keep the cows from going down the railroad right-of-way.

There was lots of green Johnsongrass along the tracks and the hungry cows started going past Ace down the tracks rather than following Charlie's pickup. Ace was running from one side of the track to the other trying to turn the cattle, but they kept going farther and farther down the tracks. When Ace stopped them on one side they would just cross the tracks and keep going down the other side.

The railroad tracks were a main route for trains going between the east and west coasts and at least a dozen long freight trains traveled them every day. It was a little after noon and as Ace waved his big hat trying to turn the cows we heard the whistle of a freight train. It was about a half a mile away and heading our way at full speed. Charlie and I weren't concerned about the cows since they were eating the grass on both sides of the track and were not likely to get in front of the train. They were used to trains going by the pasture and didn't seem spooked by the onrushing freight train.

Ace, on the other hand, was afraid that the train would hit some of the cows so he stood in the middle of the tracks frantically waving his big cowboy hat trying to stop the train. Suddenly the train's emergency brakes were applied and the 110-car train screeched to a halt a few yards from where the cows were grazing. Ace herded the cows around the engine and up the tracks to the other pasture.

About half an hour later I drove Charlie's pickup along the lane that was parallel to the tracks and the train was still stopped. Two engineers were walking along the road toward the back of the train. I stopped and asked them if they needed a ride. They were really mad and told me they never stopped a train for cattle on the tracks. They said the train's brakes had locked on one of the cars and wouldn't unlock. I asked them if there was anything I could do and one of the engineers said, "Yes. Keep that damn fool off the tracks."

About that time I heard the whistle of another train heading our way. The engineers had radioed that they were stopped so the second train slowed down and stopped behind the first one. In a few minutes a third train pulled to a halt. The Bruce's home was on the side of the Santa Anna mountain overlooking the railroad tracks. As we were eating supper about sundown we saw a long freight train heading east. In the next few minutes five more trains followed. It had been about five hours since Ace had stopped the first train and they all had to wait until the engineers could get the brakes released on the first one.

Ace said later that in all his years of cowboying that was the easiest stampede he ever stopped.

COWBOYS AND COMPASSES

Prince Wood and I were helping Red Nunley round up steers on the George West Ranch south of Batesville. The cattle were in a 5000-acre pasture and most of us had only been in the pasture once or twice so we didn't know it very well. The corrals and holding trap were on the north side of the pasture. When we found some steers we would drive them to the holding trap. The big pasture was very brushy and there weren't any landmarks. The first day was sunny so we didn't have any trouble finding the holding trap, using the sun to tell direction.

The next day was cloudy and the weather report was for clouds all day. I knew it would be easy to get lost in the pasture without the sun shining or wind blowing. It's not unusual for a cowboy to get disoriented when running cattle in an unfamiliar pasture. As I left the house that morning I picked up my compass that was about the size and shape of a pocket watch. It came out of a survival kit that I carried when I was with the Eighth Air Corps in Europe during World War II.

We were all riding together about mid-morning when we jumped a little bunch of steers. We had to run them a good ways in a circle to stop them. After we held them awhile they settled down and were ready to drive to the trap on the north fence. While we were sitting there on our horses holding the steers I pulled the compass out of my watch pocket and checked to see which way was north. The foreman hollered out that we would take the steers that way and pointed to the east. One of the other cowboys said, "I think you're wrong. We need to take them that way," and pointed to the south. A third cowboy agreed with the foreman. After everyone put his two cents worth in, everybody was still confused as to which direction to go.

I didn't say anything until one of the cowboys pointed to the north, although he wasn't too sure. I told the foreman I was sure we needed to go that direction and pointed north. I was pretty convincing and Prince Wood told the foreman that I was pretty good at finding the right direction, so we started moving the steers in the direction I had pointed. It was about two miles to the holding trap and as we were driving the cattle just about every cowboy had his opinion as to the direction we should be going. Prince and one other cowboy thought I might be right, all the rest thought we should go a different direction. The foreman insisted we would hit the south fence in a little while and would then have to drive the steers clear across the pasture. He said we might get to the trap in time for supper.

All the way to the trap I kept checking my compass to make sure we were heading north. The men thought I was checking my watch for the time of day. After a while we came out right at the gate to the holding trap and everyone said I was their hero. No one asked me if I had a compass so I didn't bother to mention it.

Another time I was helping Red Nunley and his cowboys catch some remnant steers on his Coyote Ranch south of Sabinal, Texas. The steers were really wild and had gotten away from the cowboys when they were rounding up the main bunch of cattle. The country was real brushy in the draws but fairly open on the low hills.

We had started early that morning and had roped several steers and necked them to trees near the ranch roads so they could be hauled out in a trailer later. I was riding by myself through a draw where the brush was real thick. Just as I got to the far side where the brush was the thickest a big red steer ran out of the thicket at full speed. It was solid red and twice as big as the ones we had been gathering out of the pasture. I jerked down my rope and took off after it.

I was riding one of my fastest horses and caught up with the steer before he got to the next draw. Just as I got close enough to rope him he started slowing down and I could tell he was getting ready to turn and fight me. I roped him as he was turning around to charge. The rope was tied to my saddle horn as usual and it took lots of maneuvering to keep the big steer from hooking my horse and to keep the rope from getting under my horse's feet. There weren't any trees to wrap the rope around so I had to keep moving.

I was hollering for help as the old steer chased me around a guajilla thicket several times. In a few minutes one of Mr. Nunley's cowboys came up and roped the steer. About that time Red and his foreman, Carl Nuckles, rode up. Carl said, "Red, that's one of those Tom East Santa Gertrudis bulls you bought about two years ago and cut and turned loose." He told me when they rounded up those steers last year they were one short and assumed it had died—either bled to death or killed by screwworms—since none of Red's cowboys had seen him since.

Red said those were the wildest cattle he had ever bought. This one evidently lay up in the heavy thicket during the day and grazed and watered at night. We dragged the big steer out close to a ranch road and necked him to a big mesquite tree. Red told me later he shipped the steer to the San Antonio Stockyards and it weighed 1700 pounds!

That afternoon I was riding with Red and Carl when we jumped a black two-year-old crossbred steer. I was riding the fastest horse and was really bearing down on the steer. Red and Carl were right behind me and we were flying. Just as I raised up in my stirrups to throw my loop my horse stepped in a hole and down we went. I threw the rope out to the side just as we hit the ground to keep from getting tangled up in it. I thought the horse and I would never quit rolling. Luckily the horse didn't roll on top of me.

As Red and Carl came riding up I heard Red tell Carl that he bet I was dead. I was in a sitting position as they rode up and my horse was just getting up. Red asked me if I was hurt and I said, "I sure am. I'm sitting on some prickly pear and the thorns really hurt". Red and Carl helped me pick out the long pear thorns and I got back on my horse and we went back to looking for steers.

A few years ago I phoned Carl Nuckles at Sabinal to congratulate him on his 90th birthday. After we had talked for a while he asked, "Jack, do you remember when you and your horse rolled half way across the Coyote Ranch?" I told him I sure did because it took me a week to get all the little pear thorns out!

During the terrible drought in the 1950s I was burning prickly pear cactus for about 500 cows near Batesville, Texas. The cactus on one ranch was about to run out so I made a pasture deal on the Gates Ranch east of Crystal City.

We took 400 of the cows to the Gates Ranch by driving them along the highway. It was summer and really hot and dry so we started driving them after lunch. We made it to the Paysinger Ranch by dark and put them in a small pasture for the night. There was no water in the pasture so we moved out early the next morning. We wanted to get the cattle to the Gates Ranch by noon where they could have water to drink.

The drought had been on for about three years and the tanks on the ranch were all dry. There was a big windmill and a large concrete storage tank in the pasture where we were going to burn pear. I had checked the storage tank the week before and it was full of water as were two water troughs connected to the tank. When I was there the windmill was pumping so much water it was running out of the overflow pipes of the storage tank.

We got to the ranch about noon. The weather was hot and the cows hadn't had any water since the morning before so they were really thirsty. When the cattle were 300 or 400 yards from the water troughs they could smell the water and started trotting. When they got to the troughs there was nothing but mud. There were several horses in the pasture and evidently one had pawed the float valve on one of the troughs and broken it. All the water had run out of the storage tank on to the ground. To make the situation worse the wind wasn't blowing so the windmill wasn't pumping water.

Burrell Day was helping me work cattle that summer. He was a high school student and a top cowhand. His family had a ranch near Loma Vista, south of Batesville. There we were with 400 very thirsty cows and no water. Something had to be done in a hurry. Burrell knew where there was a pump jack that could be mounted on a windmill and driven by a belt from a tractor. It would pump water from the well when the windmill wasn't turning. I knew where I could borrow a Ford tractor that had a pulley drive on it. These two very essential items were many miles apart so we headed out at full speed. We got the pump jack first and mounted it under the windmill, disconnected the sucker rod from the windmill and attached it to the pump jack.

We then drove about ten miles to get the tractor. I borrowed a pulley belt from another farmer but had to add a piece of belting to make it long enough. We used the standard alligator clamps to connect the pieces of belt together. The cattle had been milling around the troughs all day bawling for water. One of the most mournful sounds is a thirsty cow bawling for water. It's a continuous bawl that sounds pitiful, especially when 400 thirsty cows and calves amplify it.

We finally got everything ready to go about ten o'clock that night. I started the tractor and pulled the lever to start the belt that turned the pump jack. The belt started turning but then slipped off the pulley. We knew the tractor pulley and the pump jack pulley had to be in perfect alignment for the belt to stay on. We moved the tractor over a little but the belt slipped off again. We moved the tractor many times but the belt kept coming off. We drove pieces of pipe and two crowbars into the ground next to the belt, hoping that would hold the belt on the pulleys but that didn't work either.

I'll never forget the pitiful sound those 400 cows made all night long as we tried to get them water. We hadn't had any lunch or supper so about sunup Burrell suggested we go to his house and his mother would fix us breakfast. When we got to the Day Ranch Mrs. Day was fixing breakfast. She invited us in and as we were eating we told them about our dilemma.

Burrell's uncle, Bill Day, listened to our story and then asked me if I had cut the end of the belt perfectly square when adding the extra piece. I told him I thought I had. In a little while Mr. Day went to his barn and came back with a carpenter's square and some more alligator clamps. He told me to cut the other clamps off and be sure to cut the belt ends square. I thanked Mr. and Mrs. Day, and Burrell and I went back to the Gates Ranch. I cut the belt just as Mr. Day had told me and then we put it back on the tractor and pump jack. The cows were still milling around the troughs and bawling about not having any water for two days.

I crossed my fingers and started the tractor, holding my breath as I pulled the pulley lever. The belt started turning and the pump jack started going up and down. I heard water splashing on the bottom of the empty storage tank. The belt stayed on the pulleys and by noon all the cattle had their fill of water. We spent the rest of the day burning pear for the cattle and hoping it would rain.

THE BEST COWDOG I EVER OWNED

There is an old saying that a person has one good horse and one good dog in his lifetime. Mitzi was the best cowdog I ever owned. She was a Catahoula "leopard" cowdog I bought in 1949 from Morris Lightsey of Benchley, Texas. Mitzi was slate gray in color with a tan muzzle and tan feet. I bought her when she was a year old and had been trained for working cattle.

At the time I was managing the Valdina Ranch north of Sabinal, Texas. The next year, 1950, I leased a 10,000-acre ranch south of Batesville, Texas in partnership with Bob Woodward and Prince Wood. We ran about 600 mother cows on the ranch, which was covered with thick South Texas brush. Some of our cattle were really wild and hard to find in the brush, so we used cowdogs to find the cattle and help round them up.

It is hard for anyone who hasn't been to South Texas to imagine how thick the brush can be. There are places where you can't see 10 steps in any direction. Finding a path through the brush is sometimes impossible. Cattle learn to hide in the thickest brush and won't move until you are almost on top of them. Mitzi was the best I ever saw at finding cattle. She would track them through the thickest brush and bark at them until we could get there. Lots of cowdogs in South Texas will chase and bark at a deer or javelina when they are supposed to be hunting cattle. Not Mitzi, when she barked I knew she had found cattle. In the twelve years we worked together she never once let me down.

When we started driving the cattle to the corrals Mitzi would stay in front of the herd and lead them to the corrals. If the cattle wouldn't follow her I would call her back to stay behind the horses. If a cow broke out of the herd Mitzi would dash through the brush and bring her back. A good cowdog is worth several cowboys when you're working cattle in thick brush.

One of her best traits was staying with the cattle and continuing to bark so I could find them. The pastures on our ranch were 3,000 to 4,000 acres and if the wind was blowing in the wrong direction it would take me a while to locate Mitzi and the cattle. On several occasions Mitzi would have the cattle bayed but it would get too dark before I could find her. I would ride back in the pasture at daylight the next morning and Mitzi would still be with the cattle. She would be hoarse from barking all night and it would be a day or two before her bark was back to normal. You can't train a dog to do that, it's an inherited trait.

Back in those days screwworms were a real problem. They laid their eggs in open wounds on cattle and the larvae would hatch, infesting the animal. Almost every South Texas cowboy carried a bottle of screwworm medicine in a pouch, usually made from the laced-up top of an old boot, on his saddle. It didn't take Mitzi long to learn that I would rope and doctor any cattle that had screwworms. The worms generated a strong odor and when Mitzi smelled it she would start looking for the wormy animal. I would ride through a big herd of cattle with Mitzi and as soon as she scented screwworms she would bark at the animal, letting me know to doctor it.

One day I was riding along looking for screwworms when Mitzi started barking. I rode over and found a cow with a newborn baby calf that had screwworms in its navel. The cow was a big brindle crossbred with long horns that had been tipped (cut off) at the ends so they wouldn't be so sharp. The switch of her tail was gone, probably bitten off by a coyote when she was a newborn calf herself. She was about six years old and had an unusual disposition. She was very gentle and liked to be petted and have her back rubbed. In fact she would walk up to me in a corral so I could pet and rub her.

A few months after I bought her she had a calf. I drove the cow and her calf into the nearby corrals so I could put screwworm medicine on the calf's navel. As I caught the calf and threw it down to doctor it the cow really threw a fit. She bellowed, shook her head, ran her tongue out and pawed the ground. Even though she was doing all this she just didn't have that look in her eye that made me think she was going to hook me. Every cow has her own way of reacting to various situations and I had worked enough cattle to know that she probably wasn't going to do anything besides put on a good show of protecting her calf. I went ahead and doctored the calf and let up it up.

A year later the same cow had another calf with screwworms. Normally I would have driven the cow and calf to the corrals to doctor, but I knew from the previous year that the cow wasn't going to hook me so I decided to rope and doctor it right there in the pasture. She acted the same way as the year before and had I not known the cow I would have been sure she was going to hook me.

This time Mitzi found the cow and her calf right next to a fence that ran alongside a highway. As I rode up I hollered to get Mitzi's attention and signaled for her to move off a good ways from the cow and calf. I had trained Mitzi to move away while I was doctoring a calf because a cow doesn't like a dog near her calf. She will hook you and the dog if the dog is too close. Mitzi trotted off about 50 yards and I took the rope off my saddle. About this time I noticed a car coming down the highway. Just as I started swinging my loop the car slowed down and stopped about 15 feet from me. The car had Illinois license plates and I could see two couples in it.

I roped the calf, took the medicine out of my boot-top pouch, got off my horse and started toward the calf. I could tell the people in the car were watching closely. As I got to the calf and threw it down the cow went into a frenzy just like the year before. She was really putting on a show, bellowing, shaking her head, running her tongue out and dancing around breathing hard. I had my back to her the whole time and several times she got so close I could feel her hot breath on the back of my neck. I finished doctoring the calf, took my rope off the calf and started back to my horse. Just as I was getting on my horse I heard one of the men in the car say to the others, "That's the bravest man I ever saw."

By then Mitzi was known throughout the area as a top cowdog. I bred her to the best cowdogs and her offspring all made good cowdogs. Most of them inherited her trait of staying with cattle after they found them. I could have sold Mitzi's puppies for a lot of money but I gave them to friends who ranched in the area. Since I was leasing part of Mr. George West's ranch I gave him one of her puppies. There were three sisters named Carmichael who had a ranch about a mile from the West Ranch headquarters and ran Hereford cattle. One evening, when Mr. West's puppy was about a year old, the sisters drove to his home and told him his dog had their cattle bayed all day in a small pasture. They asked him to come and get the dog!

In 1954, when Mitzi was about five years old, we were moving our cattle off the West Ranch to the Paysinger Ranch near Batesville, Texas. It was during a terrible drought and we were out of grass so we had to burn pear cactus to feed them.

Burning pear involves singeing the thorns off the cactus leaves so cattle can eat them. We had rounded up cattle for several days and had them all in a large pasture north of the highway that ran through the ranch. The next day we were going to drive them into the corrals so we could take them to the Paysinger Ranch.

As we were driving them along the fence next to the highway a big crossbred cow jumped the fence, ran across the highway and jumped the other fence into a 3,000-acre pasture covered with thick brush. After we penned the herd my partner, Prince Wood, and I went to look for the runaway cow. We needed a cowdog to find the cow but Mitzi was due to have puppies any day and couldn't run very fast or far in her condition. We really needed to get the cow off the ranch so I told Prince we would take Mitzi and hope she could find the cow in a hurry.

Mitzi was too heavy with pups to jump in my pickup so I lifted her into the pickup and we drove down the highway to the place where the cow had jumped the fence. I lifted Mitzi out of the pickup and put her on the cow's trail. She waddled off through the brush trail barking after the cow. Prince and I went back to the ranch headquarters, loaded two horses in a trailer and headed back to the pasture. We drove through the pasture for about a mile and then stopped to listen for Mitzi. After a while we heard her baying off to our right about half a mile away. We unloaded our horses and rode to where Mitzi had the cow bayed in a big blackbrush thicket. When we got close the cow took off and we ran her about 300 yards before roping her near the ranch road. I held the cow from my horse while Prince rode back to get the pickup and trailer so we could load the cow.

About the time we got the cow loaded Mitzi came waddling through the brush. I picked Mitzi up, put her in the pickup and we headed back to the house. Early the next morning, as we were saddling up to move the cattle, I heard a noise that sounded like baby puppies. I shined my flashlight under the barn and sure enough there was Mitzi with seven new puppies. I wondered if they would be her best cowdog offspring yet, seeing as they were almost born on the trail of a wild cow.

In addition to being an exceptional cowdog Mitzi was excellent at raccoon hunting and finding wounded deer. When my wife Evelyn and I were first married we lived about half a mile from the Leona River. There were lots of raccoons along the river. They would come up to our barn, kill our chickens and get into the sacks of cottonseed cake that we fed our cattle. That's when Mitzi started catching raccoons. Almost every night Evelyn and I would get in our pickup, tell Mitzi to jump in the back and drive down to the river to hunt raccoons. It wouldn't take Mitzi long to bay a raccoon in a tree. I would shake it out of the tree for her to catch. We usually caught two or three raccoons before going home.

After a while Evelyn and I lost interest in raccoon hunting, but not Mitzi. Just before bedtime I would step out in the back yard and hear Mitzi baying a raccoon down on the river. I would have to get in my pickup, drive to the river and shake the raccoon out of the tree before I could take her back to the house. Some nights I wouldn't go out to listen and the next morning I would hear Mitzi baying. She had been there all night and I would have to go down to the river to get her. If I needed Mitzi to work cattle early in the morning I had to tie her up the night before or she would spend all night baying raccoons on the river and be too tired to work.

We moved to Crystal City in 1954 and lived next door to Woodrow Coleman. He had a black-and-tan hound that he used to hunt raccoons. Woodrow wanted to take his dog hunting with Mitzi so his dog could get more experience. One cold clear winter night Woodrow and I took the dogs to my ranch and turned them loose on Turkey Creek. In a few minutes Mitzi hit a raccoon trail and both dogs took off barking. They followed Turkey Creek out of my ranch and into the neighboring Ware Ranch. Then they turned northeast toward the Wilson Farm and bayed a raccoon in an old irrigation reservoir that had about a foot of water and lots of cattail plants in it.

When Woodrow and I got there his dog was on the dam barking toward the water. Mitzi wasn't anywhere around. I heard a noise out in the water and shined my flashlight toward the noise. I could see a small raccoon in a little bush just above the shallow water. I knew this raccoon was too young to have been the one the dogs were trailing. I didn't hear Mitzi barking so I immediately had the sinking feeling that Mitzi had been tracking the little one's mother and had followed it into the water. A big raccoon can easily drown a dog by hanging on to its head and holding it under water.

Even though the water was freezing cold I waded out near the little raccoon and reached under the water trying to feel Mitzi. I was wet almost up to my waist and really getting cold but I kept hoping I could find Mitzi before she drowned. Woodrow's hound kept barking from the dam but I couldn't find Mitzi. I finally gave up and we started back to the pickup when we heard several dogs barking at Mr. Wilson's house, about half a mile away. In a few minutes we heard three shotgun blasts and the dogs quit barking. That was kind of puzzling but I was so upset about losing Mitzi that we just went home. I knew Evelyn and our children would be heartbroken also.

The next morning we went to church and then to Carrizo Springs to eat lunch with friends. After lunch we drove back to Crystal City and pulled into our driveway. Just as we stopped Woodrow came around the corner of the house with a big smile on his face.

"Son, go around back and look in your doghouse," he told me. I walked around the house and to my surprise there was Mitzi asleep in her doghouse. Woodrow said he worried all night about Mitzi and the shotgun blasts. Early that morning he drove out to the Wilson Farm and there was Mitzi sitting on the front porch of Mr. Wilson's house with a small chain snapped to her collar. Mr. Wilson came out and told Woodrow that she was the best coon dog he had ever seen and began telling him what had happened the night before.

"Raccoons have been getting into my corncrib and eating the corn. Last night I heard a stray dog barking and then my own dogs started barking. I went out there with my flashlight and there was this strange dog barking at four raccoons on the rafters of the corncrib. I shot three of them and the fourth jumped down and ran under the floor of the crib with the dog right after it." Mr. Wilson said the floor was too low for Mitzi to get to the raccoon so she just barked at where it disappeared. Mr. Wilson said he left Mitzi there and went to his house to go to bed.

When Mr. Wilson got up about daylight he could hear Mitzi fighting the raccoon. He quickly got dressed and went out to the corncrib where Mitzi had just finished off the raccoon. She had been digging under the corncrib all night to get to the raccoon. Mitzi never gave up, whether she was hunting a wild cow or digging out a raccoon. Mr. Wilson caught Mitzi and saw that her collar had "Jack Kingsbery, Batesville, Texas" on it. We had been in Crystal City only a few months and I hadn't got around to getting a new collar made. Mr. Wilson and I hadn't met yet. He was going to drive to Batesville to try to find out who owned this wonderful raccoon dog when Woodrow showed up. We were all elated that Mitzi was alive and well.

I think Mitzi's favorite activity was catching rats. She would always go with me when I was burning pear to feed my cattle. Nearly all the big pear cactus plants had cactus rat nests under them. The cactus rats would pile sticks, cow chips and thorny tasajillo branches around the base of a pear cactus plant to make their nests. When I burned the thorns off the cactus leaves with a butane burner the nests would catch fire, the rats would run out and Mitzi would catch them. She quickly learned to stay about eight or ten feet from the cactus so the rats wouldn't have time to get back to the nest before she caught them. Several times while working cattle Mitzi would hear the roar of a neighbor's pear burner and take off to go catch rats. She would also catch every mouse she could find in the barn and around the house. Mitzi was just a born hunter.

Deer hunting is popular in South Texas and many times a hunter would wound a deer. If the wounded animal wasn't tracked down it would most likely die in a day or two. I started using Mitzi to trail wounded deer and in a short time she was really good at it. I had trained her not to chase deer while working cattle so she would only trail a deer if it had left some blood. It only took a drop or two of blood to start Mitzi after a wounded deer and even though it might quit bleeding, she would stay after it until she caught it.

I had another cowdog named Blue that I used to help Mitzi catch wounded deer. Blue was big and fast but I wouldn't turn him loose until Mitzi had the deer bayed. Blue would run to where Mitzi was baying and grab the deer by the neck or leg and hold it until I got there with the hunter. A slightly wounded deer can run for miles if a dog doesn't catch and hold it.

One morning Mr. Waymon Langley called and said he had wounded a big buck on his farm northwest of Crystal City. I took Mitzi, Blue and one of Mitzi's daughters, Lady, to Mr. Langley's farm to find the deer. Mr. Langley told me the deer wheeled and ran when he shot so I knew it wasn't wounded very badly. I put Mitzi on the trail and held Blue and Lady by their leashes, waiting for Mitzi to find the deer. In a little while we heard her barking. I could tell by the way Mitzi was barking that the deer was running fast and probably wasn't wounded very badly. I turned Blue and Lady loose and they took off toward the sound of Mitzi barking. We could tell the deer was running toward the Ware Ranch. After a while the dogs were too far away to hear.

We waited in Mr. Langley's field and about half an hour later here came the buck running north on the county road to the west of us. He had made almost a complete circle back to where we were. As he ran down the road the three dogs were strung out behind him. Blue was the fastest so he was about 30 yards behind the buck followed by Lady with Mitzi bringing up the rear. The buck jumped the fence and headed out across a 500-acre field that had just been plowed. Normally we chased wounded deer through thick South Texas brush where you couldn't see much but this time we could watch the whole race.

We jumped in the pickup to try to get ahead of the buck before he got to the Flanagan Ranch that I had leased. There was a reservoir on the far side of the big field and when the buck got to it he stopped to get a drink of water. As he drank he turned his head so he could watch the dogs coming across the field after him. We had stopped about fifty yards from the reservoir to try to get a shot at the buck but Mr. Langly couldn't shoot because the Kirk house was on the other side of the buck.

The buck kept drinking until Blue was about ten yards away, then he took off running and jumped the fence into the Flanagan Ranch. The dogs had run the buck about three miles and were thirsty so they stopped at the reservoir and drank before continuing the chase. Mr. Langley and I headed to the neighboring Von Rosenberg Ranch that I also had leased. We drove about a mile and half to the corner of the ranch and stopped to listen for the dogs. After about an hour we heard the dogs barking and could tell they were headed our way. In a little while we saw them coming across a pasture where the brush had been chained down. The buck was tiring and the dogs were right on his heels. When they were about 400 yards from us the buck wheeled and started fighting the dogs.

They weren't far from Palo Blanco Creek so Mr. Langley and I sneaked down the creek and got about 75 yards from where the buck and dogs were fighting. They were going round and round and Blue had the buck by the leg. I didn't want to get any closer for fear that the buck would take off running again. Mr. Langley had a 30.06 rifle with a scope and I told him to wait until the buck stopped turning and to aim at the neck so he wouldn't hit one of the dogs. He rested the rifle in the fork of a mesquite tree and took aim and waited until the buck stopped. He missed the first shot but broke the buck's neck with the second shot.

The buck was huge and we had a hard time dragging him back to the pickup. He had twelve-point antlers with a 26-inch spread and field dressed at 180 lbs. The bullet that wounded him had gone in the very bottom of his stomach and had just nicked an intestine. The buck would have eventually died if the dogs hadn't tracked him down and caught him.

Our children were little when we had Mitzi and I took them with me nearly every time we went after a wounded deer. When they saw someone who looked like a hunter drive up to our house they would start hollering, "We want to go. We want to go." One time I took the kids when we were after a wounded deer on the McKnight Ranch. It was at night and we had to run the deer about two miles before they finally caught him. I didn't have Blue, he had been hit by a car and killed earlier that year, and Mitzi and Lady couldn't stop and hold the deer. We ran all the way in the dark and I was worried about the kids keeping up. When we finally caught the buck I was more tired than the kids. The next time a hunter drove up they were ready to go again.

One summer when Mitzi was 13 years old I took her with me to catch three wild cows on a ranch east of Crystal City. Jose Flores, a good cowboy from the area, was with me to help rope and haul the cattle to the corral. Jose had helped me work cattle with Mitzi several times and he often commented that she was the best cowdog he'd ever seen. It didn't take long for Mitzi to find and bay the cows. The pasture was real brushy and as Jose and I rode up to where she was barking the three cows broke and ran. One went to the right and Jose took off after her and roped her. The other two went left and I took off after them. We ran a long way before I roped one, jerked her down and tied her. Mitzi caught up to me just as I was tying the cow. It was a very hot day and she was panting pretty hard. I hissed at her to go find the other cow and off she went.

I got back on my horse, coiled my rope and headed out in the direction the cow had gone. In a little while I heard Mitzi barking and rode to where she had the cow bayed. It took off again and I chased and roped it. As I tied the cow I expected Mitzi to show up but she didn't so I rode back to the corral figuring she would be there. She wasn't there either, so Jose and I got the pickup and trailer and loaded the three cows. When we got back to the corral Mitzi still hadn't shown up. In a situation like that I knew Mitzi would most likely return to where we had unloaded the horses.

I was getting worried so I told Jose to ride back where I had roped the last cow while I waited for her at the corral. If Mitzi wasn't there I told him to backtrack to where I had roped the first cow. I knew Mitzi would have been following my horses' tracks as I was running the last cow and she should be somewhere close by. In a little while Jose rode up carrying Mitzi's collar in one hand and wiping tears from his eyes with the other. He had found Mitzi dead, lying on the trail where she had followed me after finding her last cow.

It was a sad day for all of us. The thought kept going over in my mind, "Why did I send her after that last cow when she was already hot and tired?" It still bothers me that I didn't give Mitzi time to rest before I sent her after the second cow. Then I remember that Mitzi died doing what she loved, chasing wild cows through the South Texas brush.

Jack Kingsbery was born in 1922 and grew up on a ranch near Santa Anna in Coleman County, Texas. He was one of three brothers who were expert cowboys and horsemen by the time they were teenagers. Jack served in the Army Air Corps in Europe during World War II and graduated from Texas A&M College in 1949 with a degree in Animal Husbandry. He was a founding member of the National Intercollegiate Rodeo Association and a member of the Texas A&M rodeo team.

After graduating he managed the Valdina Ranch near Sabinal, Texas and in 1950 started ranching in a partnership on the George West Ranch near Batesville. In 1951 he married Evelyn Bruce, also a founder of the NIRA. They moved to Crystal City, Texas in 1954 and began their own ranching operation. He began breeding Beefmaster cattle in 1961 and is still active in Beefmaster Breeders United.

In 1963 Jack bought Mogford Industries, which became Kingsbery Manufacturing, and later pioneered the home/gun safe industry. He bred Quarter Horses and Thoroughbreds for racing and cow horses, and still rides and works cattle regularly.

Jack and Evelyn have four children and seven grandchildren. They live in Crystal City where Jack is active in local politics and agriculture-related organizations.